Springer Series in Advanced Microelectronics

Volume 56

The Springer Series in Advanced Microelectronics provides systematic information on all the topics relevant for the design, processing, and manufacturing of microelectronic devices. The books, each prepared by leading researchers or engineers in their fields, cover the basic and advanced aspects of topics such as wafer processing, materials, device design, device technologies, circuit design, VLSI implementation, and subsystem technology. The series forms a bridge between physics and engineering and the volumes will appeal to practicing engineers as well as research scientists.

More information about this series at http://www.springer.com/series/4076

Changhwan Shin

Variation-Aware Advanced CMOS Devices and SRAM

 Springer

Changhwan Shin
Department of Electrical and Computer Engineering
University of Seoul
Seoul
Republic of Korea (South Korea)

ISSN 1437-0387   ISSN 2197-6643   (electronic)
Springer Series in Advanced Microelectronics
ISBN 978-94-017-7595-3   ISBN 978-94-017-7597-7   (eBook)
DOI 10.1007/978-94-017-7597-7

Library of Congress Control Number: 2016940803

Printed on acid-free paper

This Springer imprint is published by Springer Nature
The registered company is Springer Science+Business Media B.V. Dordrecht

Contents

Chapter 1
Introduction: Barriers Preventing CMOS Device Technology from Moving Forward

1.1 Introduction

There is no doubt that advanced silicon integrated circuit (IC) technology has thoroughly changed the way we live. Human life has become much more convenient since hand-held mobile electronic devices were developed. This new era, called the era of the "hyper-connected society," has been facilitated by mobile devices that, in turn, have been built on the foundation established in the silicon technology revolution, when the semiconductor industries began to develop complementary metal oxide semiconductor (CMOS) technology on a continuous basis at the pace described in Moore's law [1]. For the last five decades, the physical size of transistors (or switches for digital computing) has continually shrunk, thereby enabling the number of transistors per unit chip area to double every two years. As a result of this successful shrinking of transistors, the semiconductor industry is able to fill IC chips with more transistors. Fortunately, the doubling of the density of transistors in ICs can be accomplished without an unreasonable number of fabrication process steps. In other words, we can fabricate twice as many transistors in a single wafer using almost the same number of fabrication process steps. This means that shrinking the physical size of transistors reduces the cost of manufacturing a transistor. Moreover, as the size of transistors becomes smaller, not only the density of transistors in the IC, but also the performance of transistors in the IC, can be improved. Using advanced CMOS technology, circuit designers can now propose or suggest multi-functional IC chips, and semiconductor companies can earn more money by selling more powerful devices. Much of the profit from the sale of many products is used for further investigation into the development of new CMOS fabrication processes in order to further shrink the physical size of transistors. This positive feedback loop is the main starting point in the successful development of advanced silicon technology.

© Springer Science+Business Media Dordrecht 2016   1
C. Shin, *Variation-Aware Advanced CMOS Devices and SRAM*,
Springer Series in Advanced Microelectronics 56,
DOI 10.1007/978-94-017-7597-7_1

Recently, energy efficiency has become as important as scaling and performance improvement in state-of-the-art CMOS devices. For mobile applications (e.g., smartphones), battery capacity is the one of the areas where urgent improvement is needed for sustainable growth of mobile applications. In fact, devices operating without any battery should be the ultimate goal. For wearable devices, this can be achieved by devices that are powered by the thermal energy of the human body. A more realistic approach to increase the battery life is to reduce the electricity usage of, and within, the IC chip. Electricity usage is the main technical issue for not only mobile applications, but also for cloud computing and servers. It is estimated that the electricity used by data centers represents approximately 1.4 % of all the electricity usage in the entire world [2]. An interesting fact is that the greater part of the electricity consumed is used for air conditioning or cooling to combat the heat generated in server farms. For instance, the Google data center located on the Southern coast of Finland uses very cold water in order to save the operating costs of air conditioning and cooling for the server farms. We can conclude that, if we do not alleviate this ever-increasing electricity usage, it is going to be a huge problem in the near future. Therefore, when developing CMOS technology (following Moore's law), fabricating transistors with low energy consumption should be considered to be as important as the achievement of high performance.

In considering the energy efficiency (or low power consumption) of CMOS devices, there are some fundamental limits in CMOS technology which are a hindrance to energy reduction. For example, (1) the Boltzmann Tyranny does not allow the subthreshold slope of CMOS devices to fall below 60 mV/decade at room temperature, and (2) there are always process-induced intrinsic random variations in CMOS devices. Although the semiconductor industry has adopted three dimensional (3-D) transistor structures (e.g., the fin-shaped field effect transistor, or FinFET) to offset the disadvantages of conventional planar bulk MOSFETs and to more ideally implement features of MOSFETs, the physical and fundamental limitations of CMOS transistors still inhibit significant energy reduction. Thus, in order to overcome these fundamental limitations, and to make more energy efficient ICs, various types of switches have been proposed, such as nanoelectromechanical (NEM) relays, atomic switches, non-charge-based switches, and so on. The main topic of this book has to do with another fundamental limit that comes from the steps in the CMOS fabrication process. Although the gate length of the transistor has been scaled down to the decananometer level, and the number of transistors in a chip has dramatically increased following Moore's law, the impact of atomistic process-induced variations that originate from each fabrication step has significantly increased. This causes device-to-device performance mismatches, and leads to an increase in the net power consumption of IC chips.

1.2 Overview of the Metal Oxide Semiconductor Field Effect Transistor (MOSFET)

1.2.1 Energy Crisis in MOSFET

In a digital computing device, MOSFETs are exploited as electronic switches. Ideally, MOSFETs turn off at gate voltages below the threshold voltage, and cannot conduct current when off, as shown in Fig. 1.1a. Once the electric field (from the gate) that is normal to the surface of the MOSFET is strong enough to turn on the MOSFET (i.e., higher than the threshold voltage), then current can flow from drain to source. The maximum amount of current that can flow is determined by the operating voltage of the MOSFET. However, in reality, the off-state leakage current is not actually zero, as shown in the drain current versus gate voltage plot with logarithmic scale in Fig. 1.1b. Although the drain current exponentially decreases under the threshold voltage (i.e., in sub-threshold range), there is still some leakage current when zero voltage is applied to the gate electrode of the MOSFET. Thus,

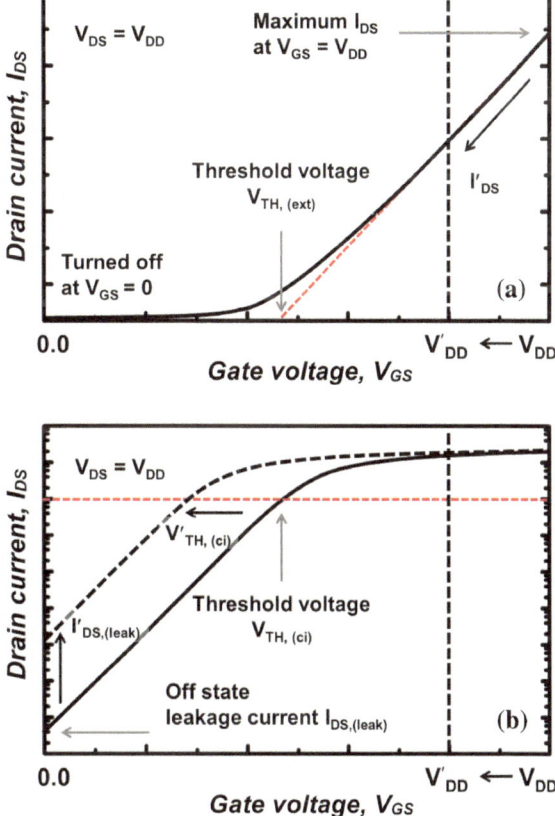

Fig. 1.1 Drain current versus gate voltage plot of a MOSFET with **a** linear scale and **b** logarithmic scale

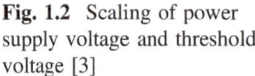
Fig. 1.2 Scaling of power supply voltage and threshold voltage [3]

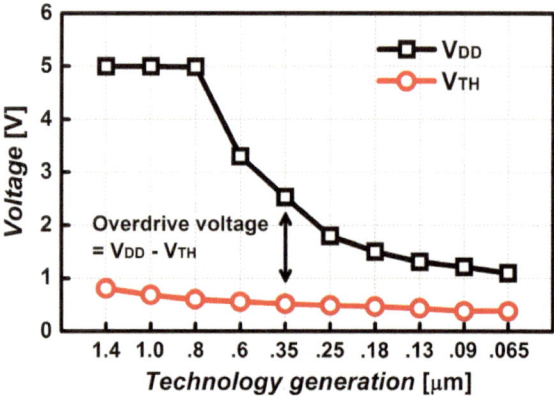

we can model the MOSFET as a switch which modulates the amount of current flow between the on-state and the off-state rather than an ideal switch which can completely cut off the current. In this situation, we need to lower the operating voltage of the transistors in order to reduce their dynamic power consumption for computing. However, if we lower the operating voltage of the transistors, the on-state drain current of the transistors also decreases (Fig. 1.1a), and this results in the degradation of computing speed and function. The only way to lower the operating voltage without degrading the on-state drain current is simply to redesign the transistor to have a lower threshold voltage (Fig. 1.1b). However, this increases off-state leakage current exponentially, leading to higher standby power consumption. Therefore, scaling down the amount of voltage swing fundamentally causes a lower on-/off-state current ratio.

For this reason, the operating voltage of CMOS chips cannot be scaled down in the same way that it is done with the physical size of transistors, because the voltage becomes saturated around 1 V, as shown in Fig. 1.2 [3]. This is because some gate overdrive (i.e., voltage difference between operating voltage and threshold voltage) needs to be maintained in order to ensure sufficient on-state drive current for circuit performance, while the threshold voltage cannot be scaled down as much as we desire because of the exponentially increasing off-state leakage current. Table 1.1 shows the power supply voltage at each semiconductor technology node, as estimated in the International Technology Roadmap for Semiconductors (ITRS). Because the power supply voltage cannot be rapidly scaled down even though the minimum pitch size of printed patterns is decreased by 0.7 × every two years, the active power density, which is consumed in switching transistors, is continually increasing, as shown in Fig. 1.3 [4]. Furthermore, because MOSFETs become

Table 1.1 Power supply voltage at each technology node (estimated by ITRS)

Technology	45 nm	32 nm	22 nm	14 nm
Supply voltage	1.0 V	0.9 V	0.8 V	0.7 V

Fig. 1.3 Increasing dynamic and leakage power density of IC [4]. Note that leakage power density dramatically increases with shortened MOSFET gate length

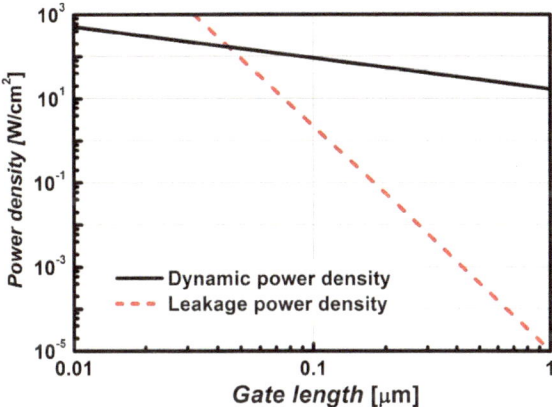

leakier, and off-state leakage current exponentially increases, the passive power density is dramatically increased as the gate length of MOSFETs is aggressively decreased. Consequently, today's IC chips also suffer from higher power density. For this reason, concepts like 'dark silicon' have emerged, which indicate regions in CMOS chips that should be turned off to allow them to cool down and thereby lower the overall temperature of the IC.

Integrated circuits based on CMOS transistors have certain limits in terms of energy efficiency. Let's consider a simple circuit which is composed of inverter chains, as shown in Fig. 1.4. During transmission of an input signal from a previous stage to a next stage, the capacitances in the circuit nodes between stages are charged and discharged. Therefore, the active power necessary to perform the functions of the circuit can be modeled as the power consumption required to charge and discharge the circuit nodes, as follows:

$$E_{\text{active}} = \alpha L_D f C V_{DD}^2 \qquad (1.2.1)$$

Fig. 1.4 An example of inverter chain

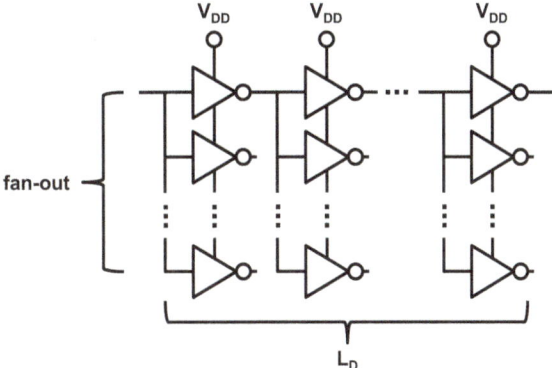

where α is activity factor, L_D is logic depth, f is fan-out, C is capacitance per stage, and V_{DD} is power supply voltage. Equation (1.2.1) explicitly shows that the active energy is proportional to V_{DD} squared, and hence, it is easily deduced that we can significantly lower the active energy by simply lowering the power supply voltage. On the other hand, ideally only one stage is switching at any point in time and the other stages are not switching (i.e. the transistors in the other stages are static). When we run the circuit, it takes some time for signals to propagate from one stage to the next stage. This delay time can be expressed as follows:

$$t_{delay} = \frac{L_D f C V_{DD}}{2 I_{ON}} \qquad (1.2.2)$$

As previously mentioned, lowering the power supply voltage inevitably decreases the on-state drive current, thereby slowing down the circuit speed (as shown in (1.2.2)). Alternatively, we can maintain the active power consumption without any degradation in delay by reducing the threshold voltage, but this necessarily increases passive power consumption (see 1.2.3).

$$E_{passive} = L_D f I_{OFF} V_{DD} t_{delay} \qquad (1.2.3)$$

$$E_{total} = E_{active} + E_{passive} = \alpha L_D f C V_{DD}^2 + L_D f I_{OFF} V_{DD} t_{delay} \qquad (1.2.4)$$

Therefore, the total amount of energy required to perform the functions of a circuit is the summation of both the active and the passive power consumptions, and this value has a fundamental minimum, as shown in Fig. 1.5. This fundamental minimum energy is currently limiting the energy efficiency of CMOS technology.

Figure 1.6a shows the total energy versus delay. As the power supply voltage (V_{DD}) is decreased, the CMOS chip consumes less energy but its delay time

Fig. 1.5 The energy required to operate CMOS circuits has a fundamental minimum value. The minimum energy point exists in the subthreshold region [5]

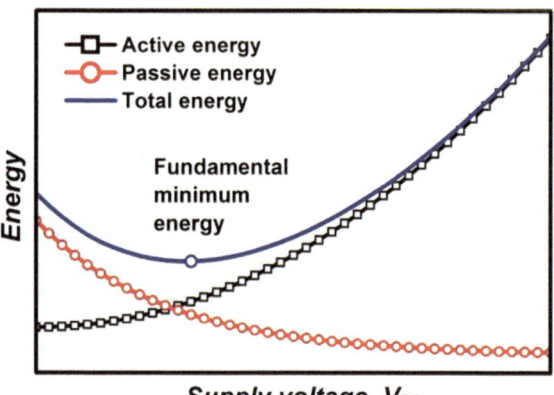

Fig. 1.6 **a** Energy versus delay plot and **b** drain current versus gate voltage plot. To increase the energy efficiency of an IC at the same delay (or to reduce the delay at the same energy), improvements are needed in the subthreshold slope of MOSFETs

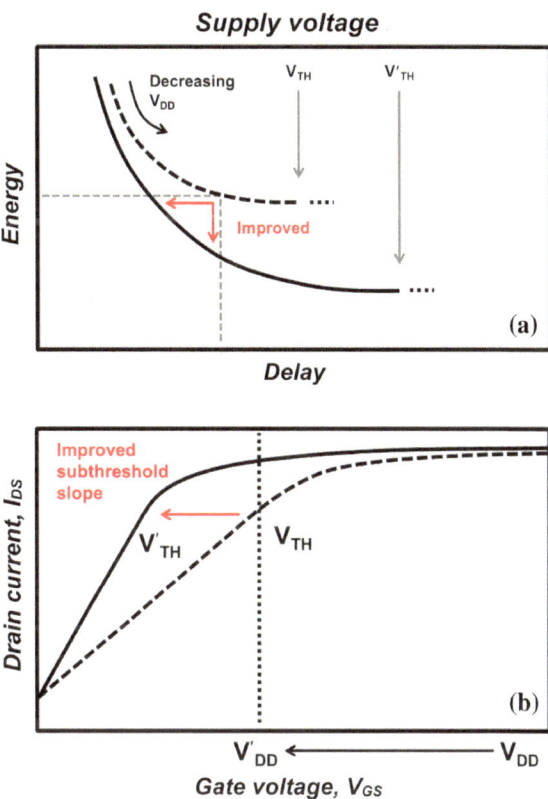

increases. If the power supply voltage is lowered below the threshold voltage, the on-/off-state current ratio starts to decrease exponentially and, at the same time, the total energy starts to increase monotonically. For this reason, it makes no sense to lower the power supply voltage beyond a certain point. In order to conserve additional energy at a given delay, we need to increase the drain current at a lower supply voltage while maintaining the off-state leakage current. This can be achieved by developing a transistor with a steeper switching characteristic, as shown in Fig. 1.6b. This transistor can conduct higher current at a lower power supply voltage, thereby making it possible to scale the power supply voltage even more before reaching the fundamental minimum. Therefore, one can conclude that improvements in the subthreshold slope (which indicates the minimum gate voltage required to increase the drain current by 10 times) in order to obtain a higher on-/off-state current ratio will allow IC chips to have better energy efficiency.

1.2.2 Thin Body MOSFETs Allow CMOS Technology to Move Forward Aggressively

The primary factor in increasing the energy efficiency of MOSFETs is a steeper subthreshold slope. However, conventional planar bulk MOSFETs have a poor subthreshold slope (e.g., SS ∼ 100 mV per decade) because the gate voltage cannot completely control the channel potential, thereby causing them to suffer from short channel effects. The value of subthreshold slope is related to the depletion-layer's capacitance and the gate oxide capacitance as follows:

$$\text{SS} = \frac{dV_G}{d\psi}\frac{d\psi}{d(\log I_{DS})} = \frac{kT}{q}\ln 10\left(1 + \frac{C_d}{C_{ox}}\right) \qquad (1.2.5)$$

where V_G is gate voltage, ψ is channel potential, I_{DS} is drain to source current, C_d is depletion layer capacitance, and C_{ox} is gate dielectric capacitance. As shown in (1.2.5), we can improve (i.e., lower) the subthreshold slope by reducing C_d or increasing C_{ox}. Therefore, the semiconductor industry has moved to utilize thin-channel or thin-body devices, such as Multigate MOSFETs (e.g., FinFET, Tri-gate MOSFET) and fully depleted silicon-on-insulator (FDSOI) MOSFETs which can lower the depletion capacitance by making the channel itself thinner [6, 7]. As shown in Fig. 1.7a, FinFETs have a three dimensional (3-D) device structure that wraps the gate electrode around the fin-shaped semiconductor channel region. Because the electric field is applied to the channel region in three different directions (i.e., front, back, and top), the gate voltage can better control the channel potential so that short channel effects are better suppressed. Because the depletion capacitance is determined by the width of the silicon channel and not by the depletion region, the subthreshold slope becomes steeper as the width of silicon channel becomes narrower. Thus, in order to achieve better gate controllability, steeper turn-on and turn-off, and a higher on-/off-state current ratio at a lower power supply voltage, the width of silicon channel should be made sufficiently narrow [8]. In general, the width of silicon channel region should be narrower than 2/3 of the gate length [8]. An n-channel FinFET with a channel length of 30 nm and a channel width of 20 nm was first demonstrated in 1998 [9]. The smallest FinFET was reported in 2006, which had a channel length of 5 nm and a channel width of 3 nm [10]. In 2011, a leader in the semiconductor industry officially announced that it would use FinFETs in mass production at the 22 nm generation of CMOS technology [11, 12]. Currently (as of 2016), the 2nd generation of 14 nm FinFET technology is in mass production [13]. Another thin body MOSFET is the FDSOI MOSFET, as shown in Fig. 1.7b. Growing a thin silicon channel on buried oxide (BOX) causes the depletion regions induced by the source/drain junction to be restricted. As a result, the FDSOI MOSFET can effectively reduce the off-state leakage current caused by punch-through and short channel effects. Furthermore, because the depletion capacitance is also limited by the thickness of the silicon

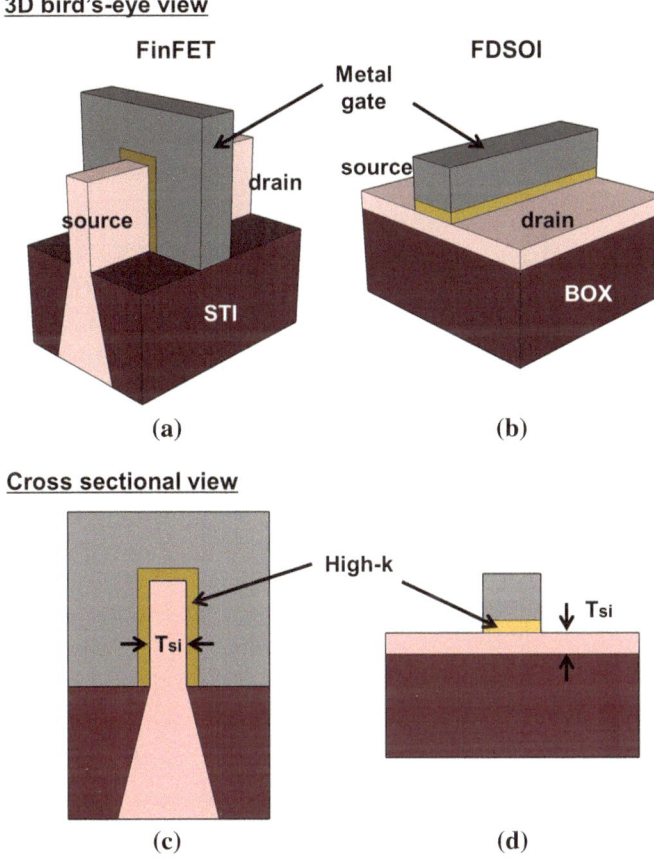

Fig. 1.7 a, b Three-dimensional (3-D) bird's-eye view of FinFET and FDSOI. **c, d** The cross sectional view of FinFET and FDSOI

channel, the FDSOI MOSFET (vs. a conventional planar bulk MOSFET) with a thin silicon channel region has a steeper subthreshold slope [7].

1.2.3 A New Class of Switch Enables CMOS Technology to Move Forward

Although thin body MOSFETs show a steeper turn-on characteristic in their input transfer characteristic curve, both the subthreshold slope of MOSFETs and the energy efficiency of CMOS chips will eventually reach limits in scalability because of the Boltzmann limit. The drain current of MOSFETs exponentially increases in the sub-threshold region because electrons in the source region are exponentially

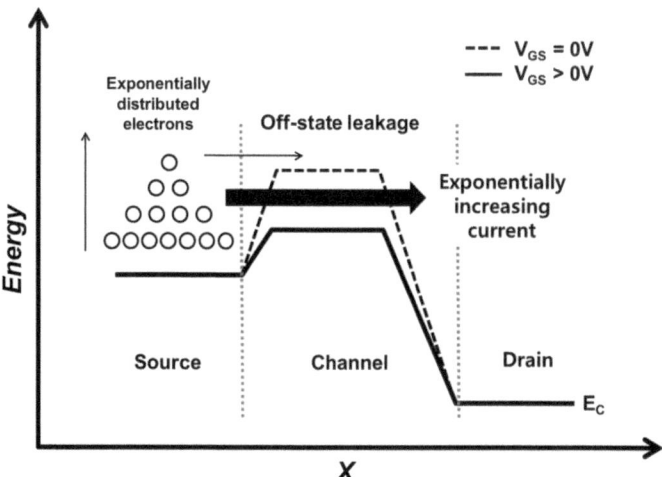

Fig. 1.8 A conceptual energy band diagram along the channel length in an n-type MOSFET. As the gate voltage increases, the number of electrons injected by the thermionic emission process is exponentially increased

distributed in energy. More specifically, when the electron energy is high enough (i.e., $E - E_F \geq 3kT$, where E_F is Fermi Energy Level), the density of electrons is exponentially decreased in accordance with the Boltzmann approximation. In the off-state, electrons that have energy higher than the height of the potential barrier at the source/channel interface, can diffuse into the channel, resulting in off-state leakage current. As the gate voltage increases, the potential barrier height decreases, and thereby, the number of electrons that can diffuse into the channel by thermionic emission process increases exponentially as shown in Fig. 1.8. Therefore, because the increment of drain current is determined by the distribution of electrons in the conduction band, even if the gate electrode completely controls the channel potential [e.g., in (1.2.5), $dV_G/d\psi = (1 + C_d/C_{ox}) = 1$], the subthreshold slope fundamentally cannot be steeper than the theoretical limit of 60 mV/decade at 300 K [i.e., in (1.2.5), $d\psi/d(\log I_{DS}) = kT\ln 10/q = 60$ mV/decade at room temperature].

In order to overcome the physical limit of the subthreshold slope, new types of switches have been proposed and studied, such as the Tunnel FET (TFET) and the Negative Capacitance FET (NCFET). These new classes of transistors adopt a different approach to solve the problem that current CMOS devices are faced with. In these new classes of devices, the limit of subthreshold slope is overcome by modifying the carrier emission process itself [i.e., $d\psi/d(\log I_{DS})$] in the TFET, and by voltage amplification [i.e., $dV_G/d\psi \leq 1$] in the NCFET. The idea behind the TFET is really to remove the exponential tail part of the carrier distribution in energy bands. As shown in Fig. 1.9, an n-type TFET uses the p-type semiconductor as a source, in contrast to the MOSFET, and there is no exponential tail because

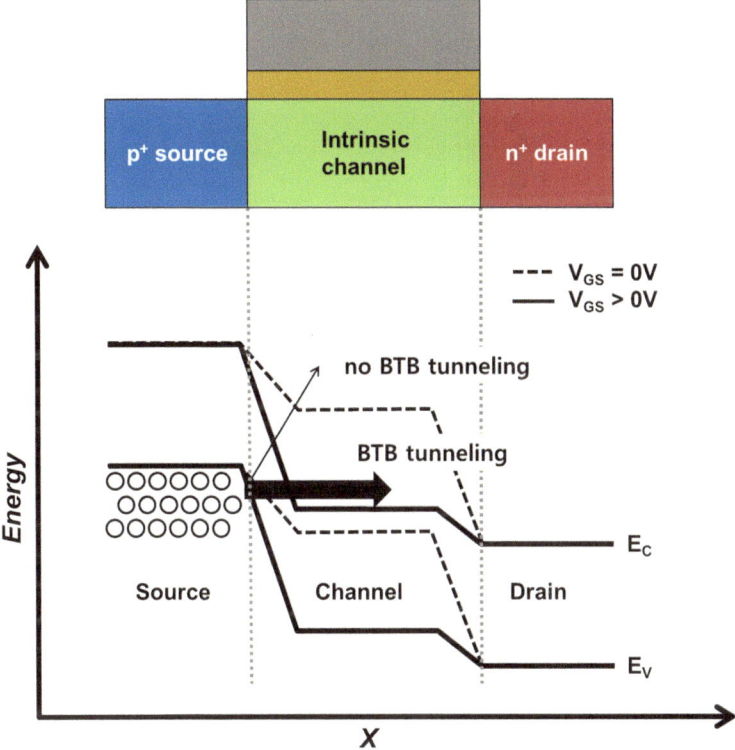

Fig. 1.9   Schematic diagram of an n-type TFET (Tunnel FET), and the energy band diagram along the channel length

there is no state in the bandgap. When the conduction band edge of the channel drops below the valence band edge of the source, lots of electrons in the source region begin to tunnel into the channel. Thus, the TFET can be turned on more abruptly and sharply, resulting in a subthreshold slope below 60 mV/decade at 300 K [14]. On the other hand, the NCFET can achieve a sub-60-mV/decade subthreshold slope while exploiting the thermionic emission process that conventional MOSFETs use. In the NCFET, a ferroelectric layer is integrated into the conventional gate stack, as shown in Fig. 1.10, and the ferroelectric capacitor has negative capacitance in a certain voltage range. When the ferroelectric capacitor is connected to a dielectric capacitor in series, the voltage drop across the dielectric capacitor is abruptly increased because the voltage across the ferroelectric capacitor is decreased in the negative capacitance region. Therefore, because the ferroelectric capacitor in the gate stack of the MOSFET acts as a voltage amplifier [i.e., $dV_G/d\psi \leq 1$], we can obtain a sub-60 mV/decade subthreshold slope in the NCFET [15].

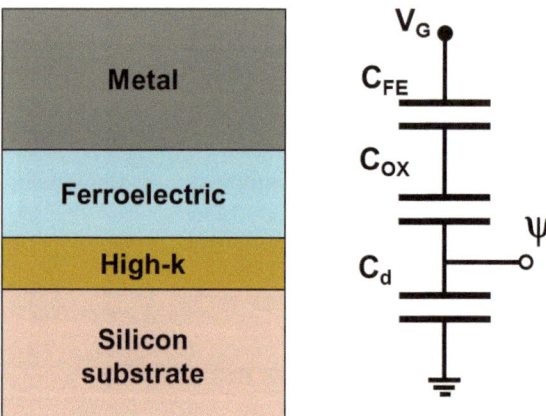

Fig. 1.10 Gate stack and equivalent circuit of the NCFET. Note that "Metal/Ferroelectric/Metal/High-k/Silicon" can be used

1.3 Process-Induced Variation

As stated in Moore's law [16], the transistor density in ICs has doubled every two years by shrinking the physical size of the transistors (or the minimum pitch, or the gate-to-gate pitch) by 30 % for each successive new CMOS technology node. As a result of this aggressive scaling, hundreds of millions of transistors (even beyond that for some applications) are integrated into today's very large scale integration (VLSI) chips, and consequently, the performance of IC chips is greatly improved at greatly reduced cost. One problem in shrinking the size of transistors down is that the power density of CMOS chips exponentially increases along with increasing transistor density. This is almost the same power density as that of a nuclear reactor [17]. In order to successfully reduce the power density of CMOS chips without undesirable degradations in the delay and yield, the characteristics of entire transistors used in IC chips should be identical. However, because process-induced systematic and random variations cause performance mismatches between transistors, power supply voltage scaling leads to unexpected degradations in delay and yield. Furthermore, as the minimum feature size of transistors is decreased to the nanometer scale, the ever-increasing impact of process-induced systematic and random variations on transistor performance is a major technical challenge that must be addressed in saving energy in CMOS chips, and will eventually adversely affect the continued scaling of CMOS technology [18–20]. Examples include the technical challenges currently being faced in scaling SRAM bit-cell area: (i) to increase the SRAM array density in ICs, (ii) to reduce the operating voltage for lower standby power consumption and longer battery lifetime in mobile/portable electronic devices, and (iii) to enhance the yield in enlarged SRAM arrays (i.e., embedded level-2 or level-3 cache memory in microprocessor/digital signal processor/system-on-chip). Therefore, the impact of process-induced systematic/random variation on the performance of transistors should be quantitatively estimated and managed in order to meet required targets for power density, performance, and yield, and ultimately to develop new CMOS technology at sub-10 nm nodes.

1.3.1   Process-Induced Systematic Variation

Process-induced variation can be divided into two parts: systematic variation and random variation. Systematic variation, including lithography proximity effects (LPE) and layout dependent strain and well proximity effects (WPE), originates from the correlation between adjacent structures. In other words, the systematic variation can occur when two (device) structures are located in close proximity to each other. Because the amount of systematic variation depends on layout, it can be predicted and modeled. Therefore, various techniques have already been developed and implemented in CMOS fabrication processes in order to effectively suppress the impact of process-induced systematic variation. For example, lithography proximity effects (which refers to a phenomenon in which mask patterns are transferred to a wafer with some degree of distortion because of the light diffraction) are compensated by improved photolithography techniques, such as optical proximity correction (OPC) and phase shift mask (PSM). Dummy structures, or patterns, are introduced in layouts to study and consider the stress effect on the transistor in order to boost its performance, and to ensure that patterning is of high quality. Furthermore, in order to suppress the process-induced systematic variation, semiconductor industries have started to design layouts in accordance with design for manufacturability (DFM) principles, including minimum poly gate pitch, poly extension length, active region length, and shallow trench isolation (STI) distance to active region, etc. Therefore, although there are many systematic variations, they do not pose an insurmountable hurdle.

1.3.2   Process-Induced Random Variation

As the channel, or active region, area of MOSFETs become smaller, device performance variability tends to be dominated by random components rather than systematic components. The performance of transistors fluctuates smoothly at low frequency due to systematic variations because neighboring devices show similar characteristics due to the close correlation between adjacent structures [21]. However, systematic variation is able to be controlled through various processes and lithography techniques, and the amount of systematic variation can be relatively negligible compared to that of random variation. In contrast, performance variation due to random variation is not predictable at all because random variation is independent of layout and exhibits no correlation between neighboring devices (i.e., performance fluctuations with high frequency) [21]. Thus, the amount of process-induced random variation can only be estimated stochastically (e.g. by characterizing the random variation in a unit of standard deviation). It is hard to completely control the amount of random variation, and transistors become more sensitive to random variation as the size of transistors is shrunk. Also, because random variation arises intrinsically from semiconductor processing, random variations are also commonly referred to as intrinsic variations. Typical examples of random variations are line edge roughness (LER) [22] in lithography steps, random dopant fluctuation (RDF) [23] in ion implantation steps, and work function variation (WFV) [24] in metal deposition steps.

The threshold voltage of a transistor is shifted by random variation, and this leads to an exponential increase in standby power consumption because the off-state leakage current is exponentially increased, as shown in Fig. 1.11. In deep logic circuits, the minimum power supply voltage for an operating circuit block increases proportionally to the amount of random variation [25] (Fig. 1.12). Because there is no static noise margin (SNM) in the voltage transfer characteristics (VTC) of six-transistor (6T) static-random-access-memory (SRAM) bit cells when the power

Fig. 1.11 Performance variation induced by random variation

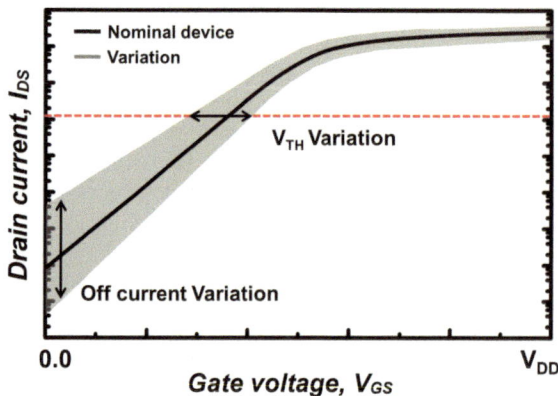

Fig. 1.12 Minimum supply voltage required to operate logic with the depth of L_D is increased with the amount of performance variation

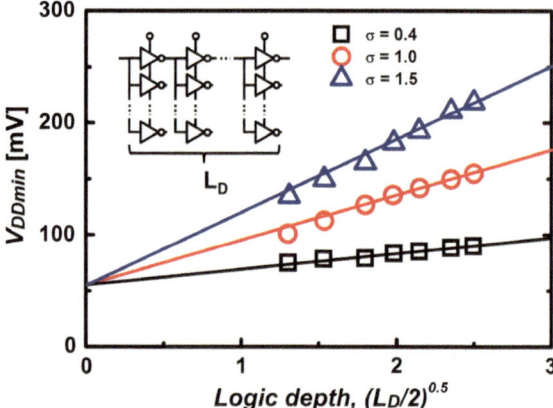

Fig. 1.13 Voltage transfer characteristics (VTC) of six-transistor (6T) static-random-access-memory (SRAM) with different supply voltages

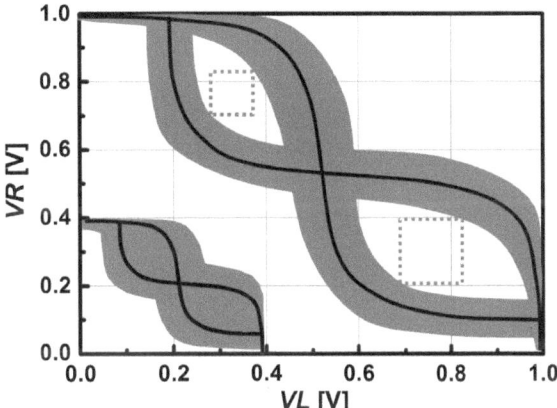

supply voltage is scaled down (Fig. 1.13), we cannot ensure that the SRAM will operate normally. In other words, performance variations induced by process-induced random variations significantly limit the opportunity to leverage power supply voltage scaling in order to reduce power consumption. Therefore, at sub-30 nm CMOS technology nodes, process-induced random variations are some of the biggest technical barriers that must be overcome when developing the next-generation of CMOS technology. This book covers the physical origins of LER, RDF, and WFV, and talks about how we can characterize each random variation source. We then discuss in detail each of these random variation sources in advanced device structures, including multi-gate and/or ultra-thin-body devices as opposed to the conventional planar bulk MOSFET. Finally, methods for improving SRAM read/write margins for given identical design rules are investigated when advanced device structures are adopted for cache memory applications.

References

1. Moore GE (1975) Progress in digital integrated electronics. IEDM Tech Dig 11–13
2. Koomey J (2011) Growth in data center electricity use 2005 to 2010. Analytics press, Oakland
3. Packan P (2007) Device and circuit interactions. IEEE International electron device meeting (IEDM '07) short course: performance boosters for advanced CMOS devices
4. Meyerson B (2004) Semico impact conference, Taiwan
5. Calhoun BH, Chandrakasan AP (2005) Ultra-dynamic voltage scaling (UDVS) using sub-threshold operation and local voltage dithering. IEEE J Solid-State Circuits 41(1):238–245
6. Suzuki K, Tanaka T, Tosaka Y, Horie H, Arimoto Y (1993) Scaling theory for double-gate SOI MOSFET's. IEEE Trans Electron Devices 40(12):2326–2329
7. Trivedi VP, Fossum JG (2003) Scaling fully depleted SOI CMOS. IEEE Trans Electron Devices 50(10):2095–2103
8. Yang J-W, Fossum JG (2005) On the feasibility of nano-scale triple gate CMOS transistors. IEEE Trans Electron Devices 52(6):1159–1164

9. Hisamoto D, Lee W-C, Kedzierski J, Anderson E, Takeuchi H, Asano K, King T-J, Bokor J, Hu C (1998) A folded-channel MOSFET for deep-sub-tenth micron era. IEDM Tech Dig 1032–1034

10. Lee H, Yu L-E, Ryu S-W, Han J-W, Jeon K, Jang D-Y, Kim K-H, Lee J, Kim J-H, Jeon SC, Lee GS, Oh JS, Park YC, Bae WH, Lee HM, Yang JM, Yoo JJ, Kim SI, Choi Y-K (2006) Sub-5 nm all-around gate FinFET for ultimate scaling. Symp VLSI Tech Dig 58–59

11. Auth C, Allen C, Blattner A, Bergstrom D, Brazier M, Bost M, Buehler M, Chikarmane V, Ghani T, Glassman T, Grover R, Han W, Hanken D, Hattendorf M, Hentges P, Heussner R, Hicks J, Ingerly D, Jain P, Jaloviar S, James R, Jones D, Jopling J, Joshi S, Kenyon C, Liu H, McFadden R, McIntyre B, Neirynck J, Parker C, Pipes L, Post I, Pradhan S, Prince M, Ramey S, Reynolds T, Roesler J, Sandford J, Seiple J, Smith P, Thomas C, Towner D, Troeger T, Weber C, Yashar P, Zawadzki K, Mistry K (2012) A 22 nm high performance and low-power CMOS technology featuring fully-depleted tri-Gate transistors, self-aligned contacts and high density MIM capacitors. Symp VLSI Tech Dig 131–132

12. Bohr M (2012) Silicon technology leadership for the mobility era. Intel Des Forum

13. Natarajan S, Agostinelli M, Akbar S, Bost M, Bowonder A, Chikarmane V, Chouksey S, Dasgupta A, Fischer K, Fu Q, Ghani T, Giles M, Govindaraju S, Grover R, Han W, Hanken D, Haralson E, Haran M, Heckscher M, Heussner R, Jain P, James R, Jhaveri R, Jin I, Kam H, Karl E, Kenyon C, Liu M, Luo Y, Mehandru R, Morarka S, Neiberg L, Packan P, Paliwal A, Parker C, Patel P, Patel R, Pelto C, Pipes L, Plekhanov P, Prince M, Rajamani S, Sandford J, Sell B, Sivakumar S, Smith P, Song B, Tone K, Troeger T, Wiedemer J, Yang M, Zhang K (2014) A 14 nm logic technology featuring 2nd-generation FinFET transistors, air-gapped interconnects, self-aligned double patterning and a 0.0588 µm² SRAM cell size. Proc IEEE IEDM 3.7.1–3.7.3

14. Choi WY, Park B-G, Lee JD, Liu T-JK (2007) Tunneling field-effect transistors (TFETs) with subthreshold swing (SS) less than 60 mV/dec. IEEE Electron Devices Lett 28(8):743–745

15. Jo J, Choi WY, Park J-D, Shim JW, Yu H-Y, Shin C (2015) Negative capacitance in organic/ferroelectric capacitor to implement steep switching MOS devices. Nano Lett 15 (7):4553–4556

16. Moore GE (1998) Cramming more components onto integrated circuits. Proc IEEE 86(1): 82–85

17. Intel® Processors—Specifications [Online] Available: http://ark.intel.com/Default.aspx

18. Sun X, Lu Q, Moroz V, Takeuchi H, Gebara G, Wetzel J, Ikeda S, Shin C, Liu T-JK (2008) Tri-gate bulk MOSFET design for CMOS scaling to the end of the roadmap. IEEE Electron Device Lett 29(5):491–493

19. Pelgrom MJM, Duinmaijer A, Welbers A (1989) Matching properties of MOS transistors. IEEE J Solid-State Circuits 24(5):1433–1440

20. Kuhn KJ (2007) Reducing variation in advanced logic technologies: approaches to process and design for manufacturability of nanoscale CMOS. Proc IEEE IEDM 471–474

21. Hiramoto T (2011) Device variability benchmark for bulk and FDSOI MOSFETs, FD-SOI Workshop, Taiwan

22. Asenov A, Kaya S, Brown AR (2003) Intrinsic parameter fluctuations in decananometer MOSFETs introduced by gate line edge roughness. IEEE Trans Electron Devices 50(5): 1254–1260

23. Asenov A (1998) Random dopant induced threshold voltage lowering and fluctuations in sub-0.1 µm MOSFETs: a 3-D "atomistic" simulation study. IEEE Trans Electron Devices 45 (12):2505–2513

24. Brown AR, Roy G, Asenov A (2007) Poly-Si-Gate-related variability in decananometer MOSFETs with conventional architecture. IEEE Trans Electron Devices 54(11):3056–3063

25. Fuketa H, Yasufuku T, Iida S, Takamiya M, Nomura M, Shinohara H, Sakurai T (2011) Device-circuit interactions in extremely low voltage CMOS designs. Proc IEEE IEDM 559–562

Part I
Understanding of Process-Induced Random Variation

Chapter 2
Line Edge Roughness (LER)

2.1 Introduction

As the physical dimensions of metal oxide semiconductor field effect transistors (MOSFETs), such as physical channel length and channel width, continue to shrink at the pace described in Moore's Law, photo-lithography technology has developed to meet the demand of printing aggressively scaled feature sizes. A brief history of the development of lithography techniques, from the 65 nm technology node to sub-10 nm technology nodes, is illustrated in Fig. 2.1. In 65 nm complementary metal oxide semiconductor (CMOS) technology, 30 nm logic gates and high density embedded memories are fabricated using ArF dry 193 nm lithography [1]. Immersion techniques for state-of-the-art photo-lithography technologies were first proposed in the 1980s [2]. While 157 nm lithography was postponed on account of strong pellicle and photoresist absorptions, 193 nm immersion lithography was rapidly adopted in the 45 nm CMOS fabrication process [3]. In sub-45 nm technology nodes, the resolution of a photoresist pattern can be further scaled down by using 193 nm immersion lithography with double exposure (DE) or double patterning (DP) [4]. Although extreme ultraviolet (EUV) lithography is expected to break through the 7 nm node with sub-15 nm resolution, technical issues, such as source power and particle contamination, prohibit its use in high-volume manufacturing.

Line edge roughness (LER) refers to the randomly varied edges of gate patterns, or the roughness of the printed pattern edge. As the minimum feature size is decreased below tens of nanometers, the effect of LER on MOSFET performance can no longer be neglected. The LER creates a few lucky channels (i.e., local short channels) in the channel length direction, resulting in device-to-device mismatch. For example, 2 % degradation in on-state drive current is experimentally observed in Intel's 65 nm devices where 3σ (where σ indicates standard deviation) of LER is greater than 10 % of the nominal gate critical dimension [5]. Because LER-induced variation is highly correlated with the short channel effect (SCE), SCE-robust

© Springer Science+Business Media Dordrecht 2016

C. Shin, *Variation-Aware Advanced CMOS Devices and SRAM*,
Springer Series in Advanced Microelectronics 56,
DOI 10.1007/978-94-017-7597-7_2

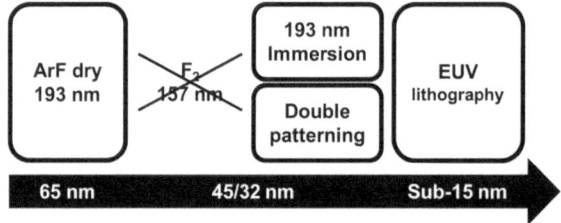

Fig. 2.1 The development history of photo-lithography techniques. For sub-30 nm CMOS technology, double patterning combined with 193 nm immersion lithography is used to fabricate extremely scaled CMOS patterns

Table 2.1 LER-induced V_{TH} variation in tri-gate MOSFETs depending on lithography technique [9]

V_{TH} variation	LER only		Total variation	
	DP (mV)	SP (mV)	DP (mV)	SP (mV)
$\sigma V_{TH, \, sat}$	12.2	15.3	49.1	49.6
$\sigma V_{TH, \, lin}$	7.0	8.6	38.5	38.7

device structures are less affected by LER-induced performance variation in a given LER profile. For example, six-transistor (6-T) SRAM cells composed of multi-gate devices, such as FinFETs and tri-gate MOSFETs [6, 7], or ultra-thin-body devices, such as FDSOI MOSFETs [8], show better immunity to LER-induced process variability because of their improved gate-to-channel capacitive coupling (in comparison with conventional planar bulk MOSFETs). Furthermore, as listed in Table 2.1, the LER-induced V_{TH} variation in 28 nm tri-gate bulk MOSFETs can be reduced by approximately 20 % by taking advantage of DP [9]. It should be noted that, despite the decrease in LER-induced V_{TH} variation, the amount of total random variation is only slightly reduced. It indicates that other random variation sources, such as random dopant fluctuation (RDF) or work function variation (WFV), are more dominant than LER in tri-gate bulk MOSFETs (note: RDF and WFV will be discussed in detail in the following sections). However, although V_{TH} variations induced by LER are reduced in FinFET devices, fin edge roughness (i.e., LER along the channel length direction) has emerged as one of the most critical random variation sources along with WFV [10]. This chapter covers (i) the root causes of LER, (ii) the method of quantitative characterization for LER profiles, and (iii) the effect of the double patterning technique on LER profiles.

2.2 Physical Origin of Line Edge Roughness

In the photo-lithography step, the pattern drawn on the mask is transferred to the resist layer because the solubility of the resist layer varies depending on whether the resist is exposed to light or not. In order to increase the sensitivity to light, chemical amplification is quite often used. In this process, chemically amplified resists are exposed to light in order to create acids. These acids then catalyze polymer deprotection during the post-exposure bake step. The deprotected portions of the resist can be easily dissolved with developer, thereby producing the resist pattern. The final LER profile contains all of the accumulated variations of each preceding processing step. In the following sections, the physical origins of LER will be introduced, followed by a discussion of each.

2.2.1 LER of Mask Patterns

When considering LER, any roughness in mask patterns would appear to be a root-cause of LER. If mask patterns themselves have LER, and lithography techniques are able to transfer that LER without distortion, the projected patterns on the resist layer will have the identical LER profile of the mask patterns. In reality, fluctuations in the mask edge are unavoidable and the mask patterns themselves have roughness. However, the amount of roughness present is small enough to neglect when compared with the original pattern size. It is technically impossible to transfer minute patterns (i.e., roughness of the mask patterns) in current 193 nm lithography. Thus, the LER inherent in the mask patterns cannot contribute to the LER of resist patterns.

2.2.2 Variations in the Dose of Light Exposure

The resolution achieved in lithography techniques primarily depends on the size of projection lens used, because the aperture (or diameter) of the projection lens determines the diffraction order. Essentially, a lens with infinite size is required to collect all diffraction orders; however, an actual lens has a finite size. This reality tends to limit the resolution of lithography techniques. The consequence is that the shape of the exposure light intensity that arrives at the resist surface is not in the shape of a step function, but rather, the shape of a sinc function (i.e., the intensity of the exposure light has a certain gradient) (Fig. 2.2). We assume that, if the intensity of exposure light is equal to the threshold intensity or higher, the resist deprotection is activated by acids and can then be easily dissolved out. The edge of the resist pattern is the point where the intensity of the exposure light is identical to the threshold dose. In order to quantitatively understand the aerial image contrast at the edge of the feature, the image log-slope (ILS) is introduced:

$$\text{ILS} = \frac{1}{I_{Edge}} \frac{\partial I(x)}{\partial x}\bigg|_{Edge},$$

Fig. 2.2 Schematic for the
photo-lithography step. Note
that the pattern edges are not
exactly matched to the mask
edges because of the gradient
of light intensity

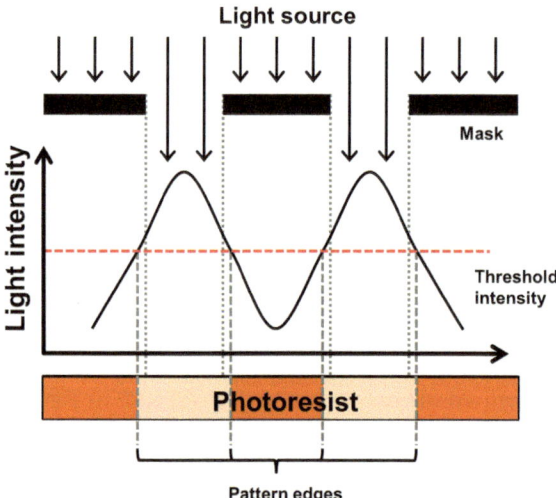

where I_{Edge} and $\frac{\partial I(x)}{\partial x}\Big|_{Edge}$ are the light intensity at the edge and the intensity slope at
the edge, respectively [11].

As each step in the lithography process is completed during fabrication, the
intensity of the exposure light in each step tends to fluctuate due to undesirable
effects, such as variations in the laser's output power, vibrations in the optical system,
miniscule up-and-down movements in the wafer stage, and/or fluctuations in the total
dose due to light quantization. Because the edges of resist patterns are determined by
the light intensity, fluctuations in exposure light intensity are one of the root-causes of
LER (Fig. 2.2). If the slope of the light intensity at the edge of a pattern is steeper, the
fluctuation of the edge is decreased. Thus, a large contrast between light and dark (i.e.,
a steep gradient of light intensity) is required in order to alleviate LER. The decrease
in LER with increasing aerial image contrast has been experimentally observed [12].
It is worth noting that, even if the aerial image contrast is continuously increased, the
LER becomes saturated at 5 nm. Beyond this point, any residual LER comes from the
intrinsic material roughness of the resist [13].

2.2.3   LER Generation in Chemically Amplified (CA) Resists

In CMOS fabrication, chemical amplification is exploited to increase the sensitivity
of the photoresist. Chemically amplified photoresist contains the photoacid generators shown in Fig. 2.3. When the photoacid generators in a chemically amplified
photoresist film absorb energy from the light, they are decomposed into acid cations
and other anions [14]. This decomposition process is referred to as deprotection.
During the post-exposure bake step, the generated acids diffuse within the resist
film and help to catalyze deprotection reactions [14, 15]. The acids are not

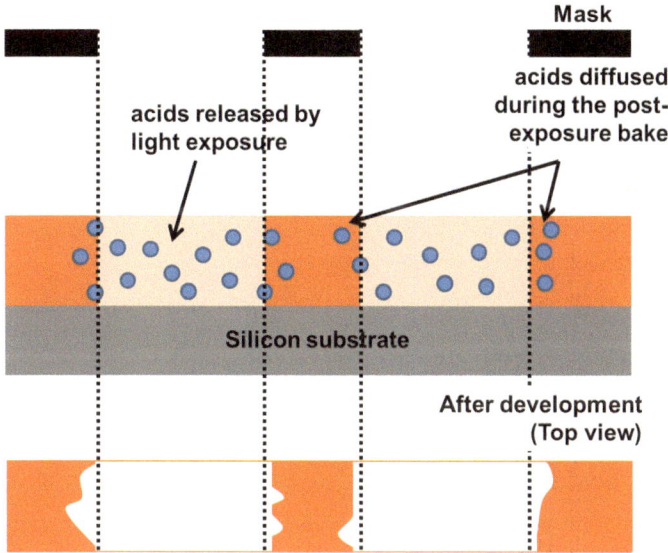

Fig. 2.3 LER, in chemically amplified resists, is formed because of acid diffusion during the post-exposure bake step

consumed but continue to exist within the deprotection reactions, and therefore are able to repeatedly catalyze the reactions. Because the acids change the solubility of polymer, the deprotected polymer regions are selectively removed with developer so that the patterns on the mask can be transferred to the resist film.

However, since the acids are randomly diffused within the resist films, this causes LER. During the post-exposure bake step, (i) bake temperature, (ii) the local extent of the deprotection reaction, and (iii) the concentration of reaction byproducts have an effect on the diffusion coefficient of the acid [16]. The diffusion distance of acid molecules is several tens of nanometers [17]. However, it is very difficult to completely control the diffusion rate because the temperature, the local extent of the reaction, and the concentration of byproducts are not constant over the baking process. Therefore, it is possible that some acids will diffuse over the target edge. If unexposed regions are sufficiently deprotected by these acids, they will be dissolved by developer, thereby causing higher frequency components in the LER (Fig. 2.3) to increase.

2.2.4   Intrinsic Roughness of the Resist

Even though other sources of LER can be excluded, intrinsic non-uniformities in photoresists cause LER along the side edges [18]. For instance, even if there is no variation in exposure light intensity, the photon absorption of photoresists varies

Fig. 2.4 Schematic of LER
due to various sizes in the
polymer chains of a
photoresist

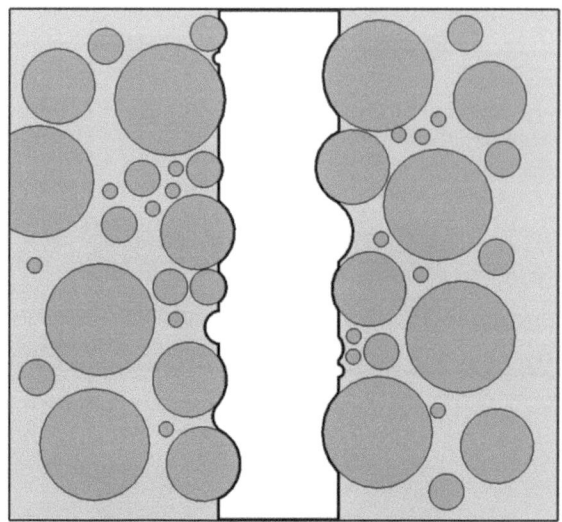

with physical position. Assuming uniform photon absorption, random dissolution and different sizes in the polymer chains of photoresists produces the roughness at the edge of the pattern (see Fig. 2.4). Furthermore, along the edges of the exposed patterns, some of the polymer molecules can smear into the developer while remaining anchored to the rest of the resist film. During the de-ionized water rinse, these partially dissolved polymer chains are re-deposited on the resist and redefine the edges of the patterns [19].

2.3 Characterization of Line Edge Roughness

2.3.1 Line Edge Roughness (LER)

LER can be measured by high resolution critical dimension scanning electron microscopes (CD-SEMs). In order to obtain the amount of LER, the local position of the line edge is first measured at regular intervals (i.e., Δ). Then, the average edge and the standard deviation of the line edge are defined as follows:

$$\bar{x} = \left(\sum_{i=1}^{N} x_i\right) \Big/ N \quad \sigma_{\text{LER}} = \sqrt{\frac{1}{N}\sum_{i=1}^{N}(\delta x_i)^2} = \sqrt{\frac{1}{N}\sum_{i=1}^{N}(x_i - \bar{x})^2},$$

where x_i is the local position measured at the ith point of the line edge. However, the average and standard deviation cannot provide us with a complete description of the LER profile because they do not include information about the spatial aspect of the LER profile (i.e., spatial frequency of the LER profile) (Fig. 2.5).

According to the self-affine edge model [20], LER and its spatial aspect can be fully described using three parameters: (i) root-mean-square (RMS) deviation (σ), (ii) correlation length (ζ), and (iii) fractal dimension (D). These three parameters can be calculated using different methods, namely, the height–height correlation function (HHCF) [21–23], the Fourier transform, or the power spectral density (PSD) [24, 25]. The correlation length (ξ) is defined as the value of the domain in which the autocorrelation function is $1/e$ or the HHCF is 1.125σ (see Fig. 2.6a) [26].

Three parameters can be calculated from the plots, namely, RMS (σ), correlation length (ξ), and fractal dimension (D). The details of each function are as follows:

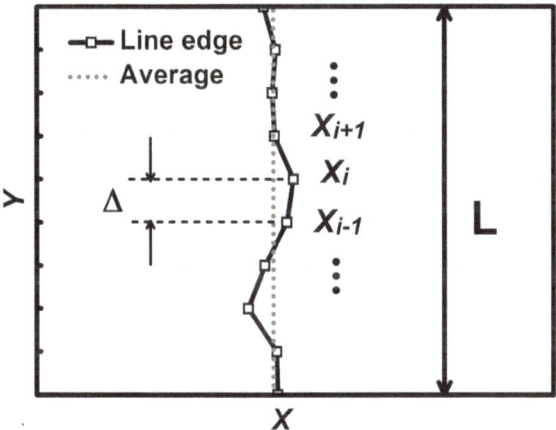

Fig. 2.5 An exemplary LER profile

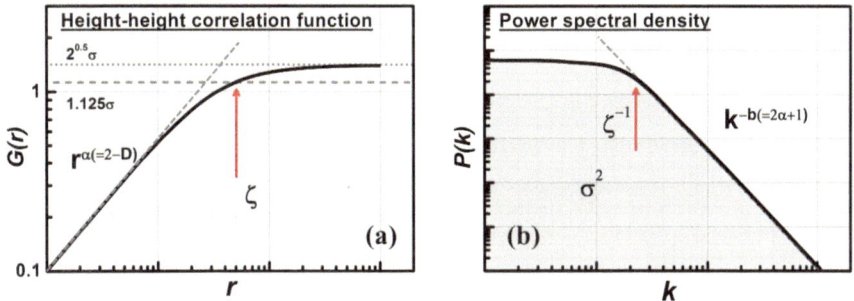

Fig. 2.6 a The height-height correlation function and **b** the power spectral density for an LER profile

$$G(md) = \left[\frac{1}{N-m} \sum_{i=1}^{N-m} (y_{i+m} - y_i) \right]^{1/2} \quad \text{(HHCF)}$$

$$R(r = md) = \frac{1}{\sigma^2} \sum_{i=1}^{N-m} (y_{i+m} - \langle y_i \rangle)(y_i - \langle y_i \rangle), \quad \text{(autocorrelation function)}$$

$$G^2(r) = 2\sigma^2[1 - R(r)],$$
(relationship between the HHCF and the autocorrelation function)

$$R(\xi) = 1/e \Rightarrow G(\xi) = \sqrt{2\left(1 - \frac{1}{e}\right)}\sigma = 1.125\sigma \quad \text{(correlation length)}$$

The correlation length means how closely the edge is correlated to its adjacent (neighboring) edge. In other words, as the value of the correlation length increases, the adjacent edge is located at a position that is similar to the position of the original edge, and the LER profile has flat hills and valleys. However, there is still a high frequency component, as shown in Fig. 2.7b. From the power spectral density (PSD), we can analyze the spatial frequency of roughness [27]. The power spectral density is obtained from the Fourier transformation of the LER profile, and provides information about the power density of spatial frequencies from 1/L to 1/Δ (where L and Δ refer to the length of the measured line and the interval distance between measurements, respectively). Therefore, we know which spatial frequency is dominant in determining the overall LER profile. Moreover, the RMS value can be obtained from the PSD using Parseval's theorem.

$$\sigma^2 = \sum_{j=1}^{N} P(k_j).$$

On the other hand, the PSD can be approximately defined as a power law in accordance with the self-affinity model.

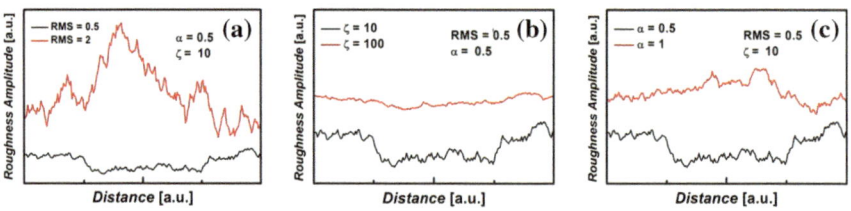

Fig. 2.7 LER profile with different **a** RMS deviation, **b** correlation length, and **c** fractal dimension (i.e., roughness exponent)

$$(P(k) \approx k^{-b}) \quad \text{for } k > \xi^{-1} \quad b = 2\alpha + 1,$$

The roughness exponent α, which is associated with fractal dimension (D) (i.e., α = 2 − D), can be extracted from the slope of the PSD using power law behavior [28, 29] (Fig. 2.6b). The fractal dimension indicates the high frequency components in the LER profile. For instance, if the value of fractal dimension is large, the slope of the power spectral density plot becomes steeper. As a result, high frequency components of the LER profile are removed, and the line edge is smoother (see Fig. 2.7c). When analyzing high frequency components of LER, the interval distance (Δ) between each sampling point in the LER profile is very important. Because the maximum measurable high frequency is determined by the interval distance (i.e., $f_{\text{Max}} = 1/\Delta$), roughness with spatial frequencies greater than f_{Max} cannot be measured. Thus, in order to use frequency sampling, an interval distance of 2 nm is recommended beyond 22 nm semiconductor technology nodes [30].

2.3.2 Line Width Roughness (LWR)

Both LER and LWR are used to estimate the amount of random variation induced by photo-lithography fabrication. LER is defined as the roughness of a single printed pattern edge, and LWR indicates the fluctuation in the physical distance between two printed pattern edges. LWR is mathematically related to LER, and can be measured using the same method that is used for LER. Assuming the measurement window covers the LWR of the gate pattern with channel length L, the width of two lines is calculated using measured positions of the left and right line at regular intervals, as follows:

$$w_i = x_i^{\text{R}} - x_i^{\text{L}},$$

where x_i^{L} and x_i^{R} are the position of the left and right edge, respectively, measured at the ith interval. Then, the average and standard deviation of line width can be calculated as follows:

$$\bar{w} = \left(\sum_{i=1}^{N} w_i \right) \bigg/ N. \quad \sigma_{\text{LWR}} = \sqrt{\frac{1}{N} \sum_{i=1}^{N} (\delta w_i)^2} = \sqrt{\frac{1}{N} \sum_{i=1}^{N} (w_i - \bar{w})^2}.$$

It can be inferred from this equation that there is a correspondence between the LWR and the LER at both the left and right edges of the resist line or layer. The standard deviation of LWR can be expressed using the standard deviation of LER at both the left and right edges as follows:

$$\sigma_{LWR}^2 = \frac{1}{N}\sum_{i=1}^{N}(w_i - \bar{w})^2 = \frac{1}{N}\sum_{i=1}^{N}\left[(w_i)^2\right] - (\bar{w})^2$$

$$= \frac{1}{N}\sum_{i=1}^{N}\left[(x_i^R)^2\right] - (\bar{x}^R)^2 + \frac{1}{N}\sum_{i=1}^{N}\left[(x_i^L)^2\right]$$

$$- (\bar{x}^L)^2 + 2\bar{x}^R \cdot \bar{x}^L - \frac{2}{N}\sum_{i=1}^{N}(x_i^R \cdot x_i^L) \qquad (2.3.1)$$

$$= (\sigma_{LER}^R)^2 + (\sigma_{LER}^L)^2 + 2\left[\bar{x}^R \cdot \bar{x}^L - \frac{1}{N}\sum_{i=1}^{N}(x_i^R \cdot x_i^L)\right],$$

where σ_{LER}^L and σ_{LER}^R are the standard deviation of LER at the left and right edge, respectively. Equation (2.3.1) explicitly shows the relationship between LWR and LER. By replacing the last term in (2.3.1) with a cross-correlation coefficient (ρ_X), (2.3.1) can be simplified, as follows:

$$\sigma_{LWR}^2 = \sigma_L^2 + \sigma_R^2 - 2\rho_X\sigma_L\sigma_R \qquad (2.3.2)$$

The value of the correlation factor depends on the method used when transferring the mask patterns, and its value is between -1 and 1. Unless additional techniques, such as double, or triple, or even quadruple patterning techniques, are used for line formation, the roughness of two edges is generally uncorrelated. When the LERs of two edges are uncorrelated (i.e., $\rho_X = 0$), there is no resemblance between them (see Fig. 2.8a). Assuming $\sigma_{LER}^L = \sigma_{LER}^R \equiv \sigma_{LER}$), the standard deviation of LWR can be written as follows:

$$\sigma_{LWR} = \sqrt{2}\sigma_{LER}^L = \sqrt{2}\sigma_{LER}^R.$$

In the case of $\rho_X = -1$, we can say that the roughness of the two edges is in anti-correlation. The two anti-correlated edges simultaneously fluctuate with opposite amplitude (see Fig. 2.8b). Thus, in the worst case, the pattern is cut off in the middle of the line. However, if the value of $\rho_X = 1$, the two edges are completely correlated, and the LERs of each edge are exactly matched (see Fig. 2.8c). Because the distance between two edges along the line is consistent to the other distances between the other two edges along the line, the standard deviation of LWR is zero ($\sigma_{LWR} = 0$).

The standard deviation of LWR provides limited information about the roughness of two line edges [26, 31]. In order to investigate spatial spectral content, Patel et al. [32] introduced a formulation of the autocorrelation function that describes the cross-correlation of a line edge with itself at different points. With regard to a stationary LWR profile, it turns out that the autocorrelation between two points is a function of the distance between them. Similarly, in a jointly stationary LWR profile, the cross-correlation coefficient in (2.3.2) is a function of the distance between them.

Fig. 2.8 LER profile of two
edges depending on the
correlation between them. For
the perfectly correlated case,
LWR is completely removed

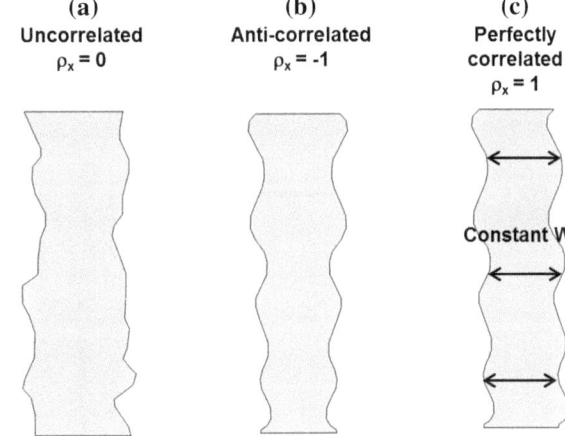

According to [33], an LWR profile can be conveniently described using the autocorrelation coefficient approximated by a closed-form expression, as follows:

$$\rho_A(y) = \exp\left[(y/\xi)^{2\alpha}\right],$$

where y is the distance between two points, ξ is the correlation length, and α is the
roughness exponent. Similar to LER, the correlation length indicates the distance
over which the amplitudes of the two points along an edge can be almost uncorrelated. The roughness exponent is a relative measure of the high-frequency
components in the roughness. Larger values mean fewer high-frequency amplitude
variations. Figure 2.7 shows the impact of each parameter on roughness.

2.4 Impact of Double Patterning on Line Edge Roughness

2.4.1 Double Pattern and Double Etching

In order to enhance the resolution of the photoresist pattern without replacing the
light source (e.g., from 193 nm to EUV), the double patterning technique was
added to the lithography process for sub-32 nm nodes [9]. The double patterning
technique has been widely adopted in industry for 22/20 nm technology and
beyond. Note that the sequence of double patterning and double etching (2P2E) is
an example of a double patterning technique. A comparison of the process flow
between double patterning and double etching versus that of conventional patterning is shown in Fig. 2.9. In the double patterning technique, the Si-BARC
and SOC are first coated onto the substrate. These layers preserve the original
pattern through the 1st and 2nd lithography steps, and play a key role as a hard
mask in the 2nd etching step. A photoresist layer is spin-coated onto Si-BARC

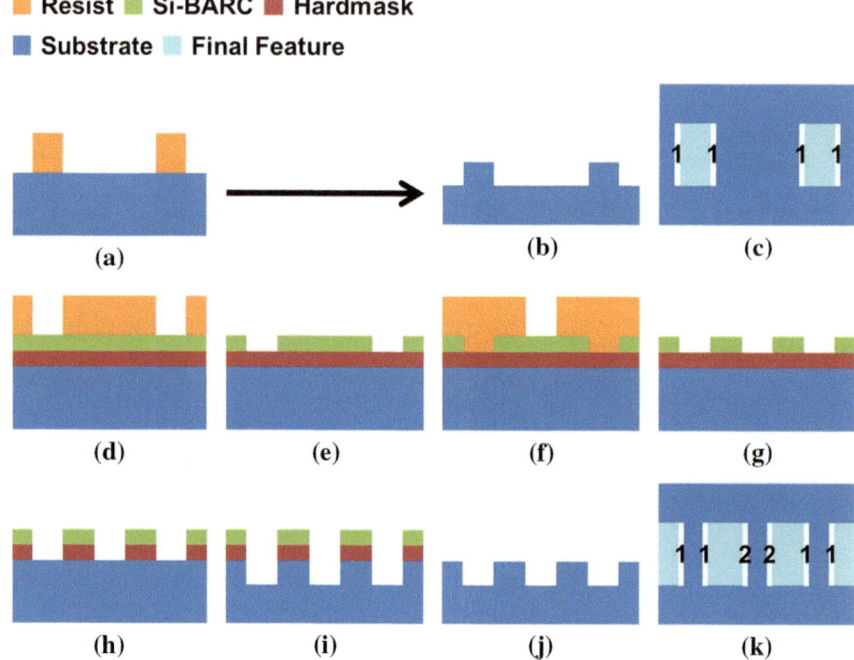

Fig. 2.9 Sequences of patterning processes for **a, b** 1P1E technique, and **d–j** 2P2E technique. Top view of the final feature for 1P1E and 2P2E is shown in (**c**) and (**k**), respectively. Note that each line edge is denoted by "1" ("2") to indicate that the line edge was affected by the first (second) patterning step

and the 1st pattern is projected by the 1st lithography process step (Fig. 2.9d). By the subsequent 1st etching step, the pattern on the resist is transferred to the Si-BARC, whereas the underlying substrate is etched in the conventional process (Fig. 2.9e, b). Next, another photoresist layer for the 2nd lithography step is spin-coated to fill out the 1st pattern on Si-BARC, and then the 2nd lithography step is performed (Fig. 2.9f). A thinner Si-BARC film can provide minimal impact on coating uniformity issues associated with coating the 2nd resist stack over the topography. The 2nd etching step is performed to transfer the pattern on the resist to the Si-BARC layer (Fig. 2.9g). Finally, the pattern is transferred from Si-BARC to SOC, and then from SOC to substrate (Fig. 2.9h–j). As a result, double patterning and double etching achieves finer patterns than conventional lithography while using identical light sources, photoresist, pitch size, and development method.

Because the LER profile is transferred through multiple etching processes, the LER profile on the substrate is different from the original LER profile on photoresist, Si-BARC, and SOC [34–39]. As the etching process is completed, the edges of the patterns tend to be smoothed. Using statistical and experimental data, it has been confirmed that the correlation length of the LER profile, based on the double patterning and double etching technique, is larger than that of the conventional LER profile [9]. Thus, multiple etching processes induce smoother line edges with low spatial frequency and flat hills/valleys, and therefore can reduce LER [40, 41]. Furthermore, additional thermal treatment, such as post-applying bake and post-exposure bake between the 1st and the 2nd lithography, increases the correlation length of the LER profile [42].

2.4.2   Self-aligned Double Patterning

Although two separate lithography steps are required to double the resolution of photoresist patterns when using the double patterning and double etching technique, there is a totally different approach, namely, self-aligned double patterning, which requires only one exposure. Self-aligned double patterning is able to double the resolution of photoresist patterns using film deposition, etching, and CMP without additional lithography steps [43]. The process flow of self-aligned double patterning is reported in [44] (Fig. 2.10). A coated photoresist is patterned with a certain pitch (note that the pitch of the final pattern will be halved) through lithography and etching steps. Next, the pattern on the photoresist is transferred and printed on a sacrificial layer by plasma etching. Then, the sacrificial layer forms a dummy gate with the duty ratio of 1:3 (i.e., line/space = 1/3). Through the deposition of silicon nitride (Si_3N_4) and anisotropic etching, spacers are formed that have identical critical dimensions to the dummy gate (i.e., the duty ratio is 1:1). The dummy gate is eliminated by an isotropic etching step, leaving only the spacer pattern on the stacked film. Finally, using the Si_3N_4 spacers as a mask for etching, the spacer patterns are transferred and printed to the hard mask. As a result of using self-aligned double patterning, the original pitch of the photoresist is decreased by

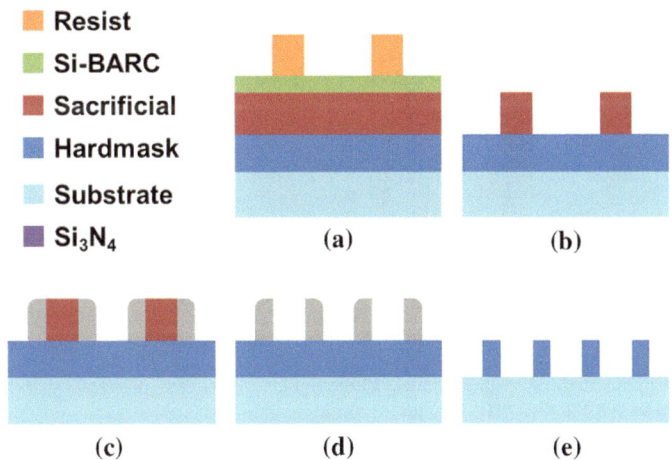

Fig. 2.10 Process flow of the self-aligned double patterning technique

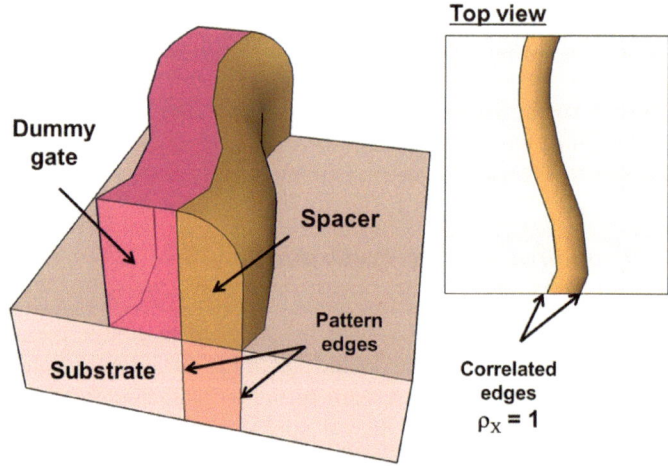

Fig. 2.11 Illustration of the self-aligned double patterning technique. This method can virtually eliminate the amount of LWR because it leads to perfectly correlated line edges

50 % in the final pattern. In other words, one resist line creates two spacers, thereby doubling the spatial frequency.

In the fabrication processes for FinFETs, the fin-shaped body can be patterned in two different ways: (1) using a resist as the mask (i.e., "resist defined"), and (2) using a spacer as the mask (i.e., "spacer-defined"). Conventional resist-defined lines create edges with uncorrelated roughness, and a ρ_X of 0 can be assumed. This is because the erosion of polymer aggregates is randomly processed for each resist edge. On the other hand, spacer-defined lines generate edges that are quite correlated. This is because of a conformal thin-film deposition process followed by a highly uniform anisotropic etch process. These preceding steps induce a spacer mask to be formed along the sidewall of a dummy resist-defined feature (Fig. 2.11). If the spacer width (corresponding to the thickness of the deposited film) is negligible (versus the inverse value of the LWR spatial frequency), the spacer-defined lines have a uniform width. Hence, a ρ_X of 1 can be assumed. In summary, if the self-aligned double patterning technique is used in the FinFET fabrication process, the performance variation induced by LWR (not LER) can be virtually eliminated.

References

1. Hashimoto K, Uesawa F, Takahata K, Kikuchi K, Kanai H, Shimizu H, Shiobara E, Takeuchi K, Endo A, Harakawa H, Miniotogi S (2003) ArF lithography technologies for 65 nm-node CMOS (CMOSS) with 30 nm logic gate and high density embedded memories. In: Symposium on VLSI Technology Digest, pp 45–46
2. Lin BJ (1987) The future of subhalf-micrometer optical lithography. Microcircuit Eng 6 (1):31–51

3. Narasimha S, Onishi K, Nayfeh HM, Waite A, Weybright M, Johnson J, Fonseca C, Corliss D, Robinson C, Crouse M, Yang D, Wu C-HJ, Gabor A, Adam T, Ahsan I, Belyansky M, Black L, Butt S, Cheng J, Chou A, Costrini G, Dimitrakopoulos C, Domenicucci A, Fisher P, Frye A, Gates S, Greco S, Grunow S, Hargrove M, Holt J, Jeng S-J, Kelling M, Kim B, Landers W, Larosa G, Lea D, Lee MH, Liu X, Lustig N, McKnight A, Nicholson L, Nielsen D, Nummy K, Ontalus V, Ouyang C, Ouyang X, Prindle C, Pal R, Rausch W, Restaino D, Sheraw C, Sim J, Simon A, Standaert T, Sung CY, Tabakman K, Tian C, Van Den Nieuwenhuizen R, Van Meer H, Vayshenker A, Wehella-Gamage D, Werking J, Wong RC, Yu J, Wu S, Augur R, Brown D, Chen X, Edelstein D, Grill A, Khare M, Li Y, Luning S, Norum J, Sankaran S, Schepis D, Wachnik R, Wise R, Wann C, Ivers T, Agnello P (2006) High performance 45-nm SOI technology with enhanced strain, porous low-k BEOL, and immersion lithography. In: Proceedings of IEEE IEDM, pp 1–4
4. Chen H-Y, Chang C-Y, Huang C-C, Chung T-X, Liu S-D, Hwang J-R, Liu Y-H, Chou Y-J, Wu H-J, Shu K-C, Huang C-K, You J-W, Shin J-J, Chen C-K, Lin C-H, Hsu J-W, Perng B-C, Tsai P-Y, Chen C-C, Shieh J-H, Tao H-J, Chen S-C, Gau T-S, Yang F-L (2005) Novel 20 nm hybrid SOI/Bulk CMOS technology with 0.183 μm^2 6T-SRAM cell by immersion lithography. In Symposium on VLSI Technology Digest, pp 16–17
5. Chandhok M, Datta S, Lionberger D, Vesecky S (2007) Impact of line width roughness on Intel's 65 nm process devices. In: Proceedings of SPIE, p 65191A
6. Shin C, Damrongplasit N, Sun X, Liu T-JK (2011) Performance and yield benefits of quasi-planar bulk CMOS technology for 6-T SRAM at the 22-nm node. IEEE Trans Electron Devices 58(7):1846–1854
7. Shin C, Tsai CH, Wu MH, Chang CF, Liu YR, Kao CY, Lin GS, Chiu KL, Fu C-S, Tsai C, Liang CW, Nikolić B, Liu T-JK (2011) Quasi-planar bulk CMOS technology for improved SRAM scalability. Solid-State Electron 65–66:184–190
8. Shin C, Cho MH, Tsukamoto Y, Nguyen B-Y, Mazuré C, Nikolić B, Liu T-JK (2010) Performance and area scaling benefits of FD-SOI technology for 6-T SRAM cells at the 22-nm node. IEEE Trans Electron Devices 57(6):1301–1309
9. Shin C, Park IJ (2013) Impact of using double-patterning versus single patterning on threshold voltage (V$_{TH}$) variation in quasi-planar tri-gate bulk MOSFETs. IEEE Electron Device Lett 34 (5):578–580
10. Wang X, Brown AR, Cheng B, Asenov A (2011) Statistical variability and reliability in nanoscale FinFETs. In: Proceedings of IEEE IEDM, pp 5.4.1–5.4.4
11. Wei Y, Brainard RL (2009) Advanced processes for 193-nm immersion lithography. SPIE, Bellingham
12. Pawloski AR, Acheta A, Bell S, La Fontaine B, Wallow T, Levinson HJ (2006) The transfer of photoresist LER through etch. In: Proceedings of SPIE, p 615318
13. Tsubaki H, Yamanaka T, Nishiyama F, Shitabatake K (2007) A study on the material design for the reduction of LWR. In: Proceedings of SPIE, p 651918
14. Tagawa S, Nagahara S, Iwamoto T, Wakita M, Kozawa T, Yamamoto Y, Werst D, Trifunac AD (2000) Radiation and photochemistry of onium salt acid generators in chemically amplified resists. In: Proceedings of SPIE, p 204
15. Wang X-B, Ferris K, Wang L-S (2000) Photodetachment of gaseous multiply charged anions, copper phthalocyanine tetrasulfonate tetraanion: tuning molecular electronic energy levels by charging and negative electron binding. J Phys Chem A 104(1):25–33
16. Stewart MD, Tran HV, Schmid GM, Stachowiak TB, Becker DJ, Willson CG (2002) Acid catalyst mobility in resist resins. J Vac Sci Technol, B 20(6):2946–2952
17. Hinsberg WD, Houle FA, Sanchez MI, Hoffnagle JA, Wallraff GM, Medeiros DR, Gallatin GM, Cobb JL (2003) Extendibility of chemically amplified resists: another brick wall? In: Proceedings of SPIE, p 1
18. Nam H, Lee GS, Lee H, Park IJ, Shin C (2014) Analysis of random variations and variation-robust advanced device structures. J Semicond Technol Sci 14(1)

19. Prabhu VM, Vogt BD, Kang S, Rao A, Lin EK, Satij SK, Turnquest K (2007) Direct measurement of the in situ developed latent image: the residual swelling fraction. In: Proceedings of SPIE, p 651910
20. Zhao Y (2001) Characterization of amorphous and crystalline rough surface: principles and applications. Academic Press, San Diego
21. Constantoudis V, Patsis GP, Tserepi A, Gogolides E (2003) Quantification of line-edge roughness of photoresists. II. Scaling and fractal analysis and the best roughness descriptors. J Vac Sci Technol B 21:1019–2003
22. Patsis GP, Constantoudis V, Tserepi A, Gogolides E, Grozev G, Hoffmann T (2002) Roughness analysis of lithographically produced nanosturctures: off-line measurement and scaling analysis. Microelectron Eng 67–68:319–325
23. Constantoudis V, Patsis GP, Gogolides E (2003) Photoresist line-edge roughness analysis using scaling concepts. In: Proceedings of SPIE, p 901
24. Yamaguchi A, Tsuchiya R, Fukuda H, Komuro O, Kawada H, Iizumi T (2003) Characterization of line-edge roughness in resist patterns and estimations of its effect on device performance. In: Proceedings of SPIE, p 689
25. Bunday BD, Bishop M, Villarubia JS, Vladar AE (2003) CD-SEM measurement line-edge roughness test patterns for 193-nm lithography. In Proceedings of SPIE, p 674
26. Constantoudis V, Patsis GP, Leunissen LHA, Gogolides E (2004) Line edge roughness and critical dimension variation: fractal characterization and comparison using model functions. J Vac Sci Technol B 22(4):1974–1981
27. Naulleau PP, Cain JP (2007) Experimental and model-based study of the robustness of line-edge roughness metric extraction in the presence of noise. J Vac Sci Technol B 25 (5):1647–1657
28. Barabasi A-L, Stanley HE (1995) Fractal concepts in surface growth. Cambridge University Press, Cambridge
29. Zhao BY, Wang G-C, Lu T-M (2001) Characterization of amorphous and crystalline rough surface: principles and applications experimental methods in the physical sciences academic, New York
30. Bunday BD, Bishop M, McCormack D, Villarrubia JS, Vladar AE, Dixson R, Vorburger T, Orji NG, Allgair JA (2004) Determination of optimal parameters for CD-SEM measurement of line edge roughness. In: Proceedings of SPIE, p 515
31. Constantoudis V, Gogolides E, Roberts J, Stowers J (2005) Characterization and modeling of line width roughness (LWR). In: Proceedings of SPIE, p 1227
32. Patel K, Liu T-JK, Spanos CJ (2009) Gate line edge roughness model for estimation of FinFET performance variability. IEEE Trans Electron Devices 56(12):3055–3063
33. Palasantzas G (1993) Roughness spectrum and surface width of self-affine fractal surfaces via the K-correlation model. Phys Rev B Condens Matter 48(19).14472–14478
34. Mahorowala P, Babich1 K, Lin Q, Medeiros DR, Petrillo K, Simons J, Angelopoulos M, Sooriyakumaran R, Hofer D, Reynolds GW, Taylor JW (2000) Transfer etching of bilayer resists in oxygen-based plasmas. J Vac Sci Technol A 18(4):1411–1419
35. Mahorowala AP, Goldfarb DL, Temple K, Petrillo KE, Pfeiffer D, Babich K, Angelopoulos M, Gallatin GM, Rasgon S, Sawin HH, Allen SD, Lang RN, Lawson MC, Kwong RW, Chen K-J, Li W, Varanasi PR, Sanchez MI, Ito H, Wallraff GM, Allen RD (2003) Impact of thin resist processes on post-etch LER. In: Proceedings of SPIE, p 213
36. Montgomery PK, Peters R, Garza C Sr, Cobb J, Darlington B, Parker C, Filipiak S, Babbitt D (2005) Reduction of line edge roughness and post resist trim pattern collapse for sub 60 nm gate patterns using gas-phase resist fluorination. In: Proceedings of SPIE, p 1024
37. Namatsu H, Nagase M, Yamaguchi T, Yamazaki K, Kurihara K (1998) Influence of edge roughness in resist patterns on etched patterns. J Vac Sci Technol B 16(6):3315–3321
38. Ren F, Pearton SJ, Lothian JR, Abernathy CR, Hobson WS (1992) Reduction of sidewall roughness during dry etching of SiO_2. J Vac Sci Technol B 10(6):2407–2411
39. Ren F, Pearton SJ, Shul RJ, Han J (1998) Improved sidewall morphology on dry-etched SiO_2 masked GaN features. J Electron Mater 27(4):175–178

40. Wallow T, Acheta A, Ma Y, Pawloski A, Bell S, Ward B, Tabery C, Fontaine BL, Kim R-H, McGowan S, Levinson HJ (2007) Line-edge roughness in 193-nm resists: lithographic aspects and etch transfer. In: Proceedings of SPIE, p 651919
41. Goldfarb DL, Mahorowala AP, Gallatin GM, Petrillo KE, Temple K, Angelopoulos M, Rasgon S, Sawin HH, Allen SD, Lawson MC, Kwong RW (2004) Effect of thin-film imaging on line edge roughness transfer to underlayers during etch processes. J Vac Sci Technol B 22 (2):647–653
42. Steenwinckel DV, Lammers JH, Leunisse LHA, Kwinten JAJM (2005) Lithographic importance of acid diffusion in chemically amplified resists. In: Proceedings o SPIE, p 269
43. Hand A (2007) Applied's litho scheme: patterning vs. printing. Semiconductor International, April 2007
44. Mukai H, Shiobara E, Takahashi S, Hashimoto K (2007) A study of CD budget in spacer patterning technology. In: Proceedings of SPIE, p 692406

Chapter 3
Random Dopant Fluctuation (RDF)

3.1 Introduction

Following Moore's Law [1], semiconductor industries have doubled the density of transistors in integrated circuits (ICs) every two years. This has rapidly increased the performance of ICs because the degree of integration has grown exponentially. However, below the 1 μm technology node, a serious technical issue was encountered that frustrated further shrinking of the gate pitch, namely, the short channel effect (SCE) [2, 3]. The short channel effect brings about other undesirable effects, such as threshold voltage (V_{TH}) roll off [4–6] and drain-induced barrier lowering (DIBL) [7–9]. As the portion of the channel region depleted by the source/drain junction increases as much as the channel length is shortened, the threshold voltage is reduced. In other words, lower gate voltage is required to invert the channel region because a larger part of the channel region is already depleted by the source/drain-to-channel junctions. The second phenomenon, namely, drain-induced barrier lowering (DIBL), occurs when the source-to-channel potential barrier is affected by the drain bias. If the gate length of metal oxide semiconductor field effect transistors (MOSFETs) is scaled down to such an extent that the energy barrier height at the source-to-channel interface is decreased because of the electric field emanating from the drain region, more electrons can diffuse from the source to the drain over the energy barrier at the source/channel interface. This causes higher off-state leakage current, degraded subthreshold slope, and a lowered threshold voltage. In order to alleviate short channel effects, the channel region of MOSFETs should be sufficiently doped to minimize (i) the depletion charge induced by the source/drain junction, and (ii) the impact of drain bias on the modulation of the height of the energy barrier at the source/channel interface [10]. In very scaled MOSFETs in sub-32 nm technology nodes, the doping concentration in the channel (or the halo doping concentration) is close to 10^{18} cm^{-3} or even higher [11, 12].

© Springer Science+Business Media Dordrecht 2016

C. Shin, *Variation-Aware Advanced CMOS Devices and SRAM*,
Springer Series in Advanced Microelectronics 56,
DOI 10.1007/978-94-017-7597-7_3

Fig. 3.1 Average number of
dopant atoms in the channel
region versus the technology
node [13]

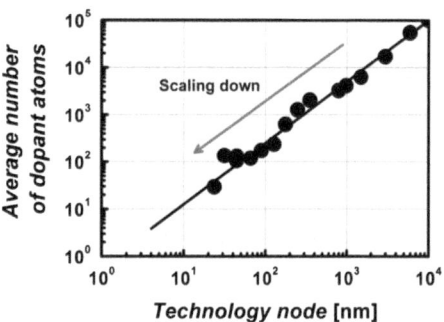

Unfortunately, a heavily-doped channel region (i.e., higher than 10^{18} cm^{-3}) causes intrinsic random variations (i.e., random dopant fluctuations) in aggressively scaled MOSFETs at sub-30 nm technology nodes and beyond. Figure 3.1 shows the average number of dopants in the channel region of MOSFETs versus CMOS technology nodes [13]. The average number of dopant atoms in the channel region of micron-scale sized MOSFETs numbers in the thousands. Thus, the effects of fluctuation in the exact number of dopants are not significant. In contrast, nano-scaled MOSFETs have only a few tens of the average number of dopant atoms in their channel region, even though the doping concentration of the channel region is increased to prevent short channel effects, as mentioned above (e.g., considering a cube with a side of 30 nm and doping concentration of 10^{18} cm^{-3}, the number of dopants in the cube is 10^{18} cm$^{-3} \times 30$ nm$^3 = 27$). This simple calculation counts the number of impurity atoms in a cube (i.e., in the channel region of transistor in volume), and implicitly indicates that device performance can be determined by fewer and fewer dopants, thereby becoming vulnerable to the fluctuation in the exact number of dopants. Even if MOSFETs have the identical number of dopants, the spatial position of dopants in the channel region can vary, resulting in device performance variation. From the analytically expressed MOSFET threshold voltage, we can estimate the impact of random dopant fluctuations on the threshold voltage. The threshold voltage of n-type MOSFETs is expressed as follows:

$$V_{TH} = V_{FB} + 2\phi_F + \frac{\sqrt{4q\varepsilon_{si}\varepsilon_o N_a \phi_F}}{C_{OX}}, \tag{3.1.1}$$

where V_{FB} is the flat band voltage, ϕ_F is the bulk potential, C_{OX} is the gate oxide capacitance, and ε_{si} and ε_o are the permittivity of silicon and vacuum, respectively. Equation (3.1) explicitly shows the relationship between the threshold voltage of the MOSFETs and the doping concentration of the MOSFETs' channel region. Based on this, we know that random dopant fluctuation definitely contributes to the threshold voltage variation of MOSFETs. In this chapter, the physical origin of random dopant fluctuation (RDF) will be addressed and characterized.

3.2 Physical Origin of Random Dopant Fluctuation

In modern IC (Integrated Circuit) fabrication, ion implantation is usually employed to dope semiconductor materials. Ion implantation is a doping process by which various ions of a particle (accelerated by an electric field) penetrate into the surface of a substrate target. Ion implantation is very useful for selectively doping and/or forming shallow junctions. Ion implantation can be divided into two parts: (i) an implantation step for injecting impurities, and (ii) an annealing step for repairing damage and activating injected impurities. Each step is involved in the root-cause analysis of the random distribution of dopants in the channel region.

3.2.1 Ion Implantation Step

During the ion implantation step, impurity source ions are accelerated in the acceleration tube as they move toward the substrate before being shot into the target wafer. Once the ions penetrate through the surface of the wafer, they are usually stopped by collision events with electrons and nuclei in the substrate, including nuclear stopping, nonlocal electronic stopping, and local electronic stopping [14]. However, some lucky electrons are able to penetrate deeply into the semiconductor substrate without undergoing any collisions (called "channeling" [15]). To block these channeling ions, an amorphous layer, such as oxide and/or damaged crystal, is deposited on the surface layer of the wafer prior to the ion implantation processing step. Another way of blocking the channeling ions is to tilt the direction of the accelerated ions with respect to the target wafer. When the incident ions collide with a stationary target atom (i.e., nuclear stopping), as shown in Fig. 3.2a, the incident ions lose their own energy and the direction of the ion's trajectory is altered by the interaction with the internal electric field of the nucleus in the target atom. Nonlocal electronic stopping is different from nuclear stopping and occurs when the incident ions travel through a dielectric medium. The polarization field results in

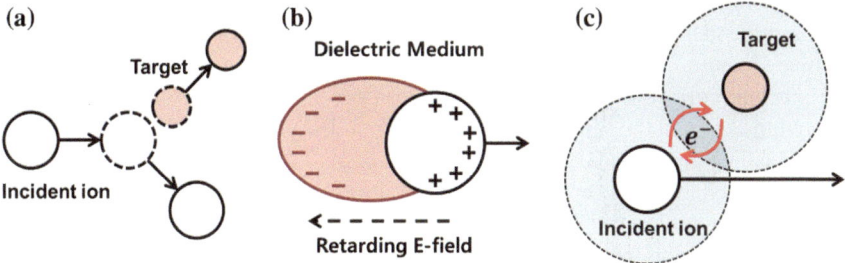

Fig. 3.2 Stopping mechanism in ion implantation: **a** nuclear stopping, **b** non-local electronic stopping, and **c** local electronic stopping [14]

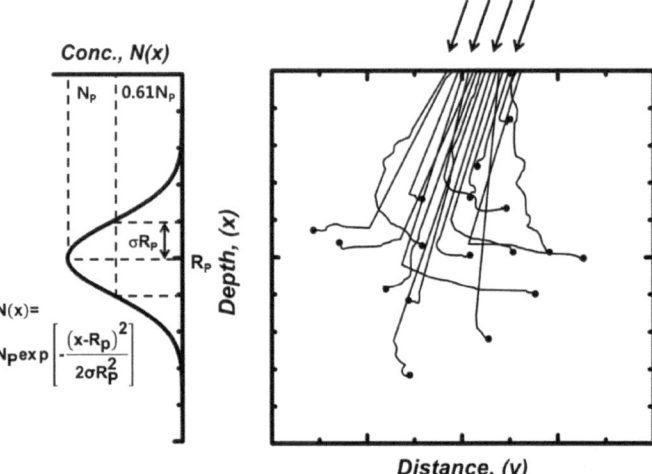

Fig. 3.3 Schematic diagram of randomly distributed dopant atoms. The distribution can be approximated by the Gaussian distribution

drag force, which decreases the energy of the incident ions but cannot change their direction, as shown in Fig. 3.2b. Finally, when the incident ions move and pass close to a lattice atom, a momentum exchange occurs, leading to deflection of the incident ions (Fig. 3.2c). In the ion implantation process, these three primary collisions cannot be completely controlled. Therefore, after these collisions, the energy and trajectory of incident ions are randomly distributed. As a result, although the distribution of implanted dopant atoms approximately follows the Gaussian distribution [15], the impurities are randomly dispersed in the substrate (see Fig. 3.3).

3.2.2 Annealing Step for Repairing Damage and Activating Impurities

After going through a series of nuclear and electronic collisions, implanted ions in the semiconductor substrate lose their energy and finally stop somewhere within the lattice structure. An interesting fact is that the stationary target atoms are displaced by nuclear collisions. In other words, because a huge amount of energy is transferred to atoms in the semiconductor, nuclear collisions cause implant damage (also called lattice disorder) [16]. However, in the case of electronic collisions, the atoms in the semiconductor keep their original positions in the lattice structure, and lattice disorder does not occur [17]. The shape of the disorder depends on material itself (Fig. 3.4) because dominant collisions and energy losses are dependent on the mass of the material (see Fig. 3.5). The displacement energy (E_d) is defined as the minimum energy required to displace a semiconductor atom out of its original

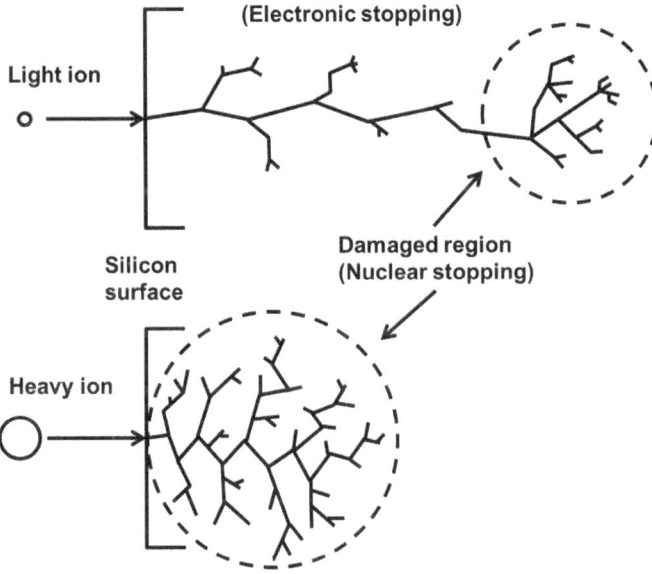

Fig. 3.4 Lattice disorder caused by ion implantation [17]. The shape of the disorder varies with the mass of dopants

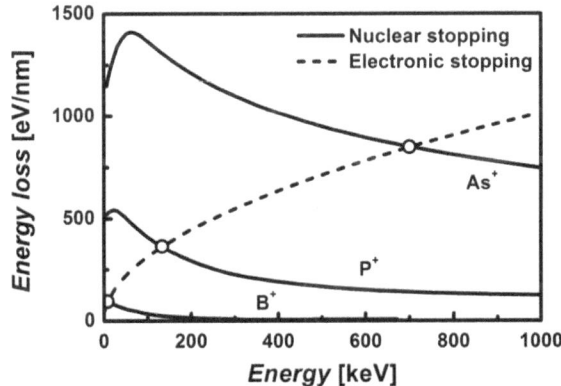

Fig. 3.5 Energy loss due to nuclear stopping and electronic stopping for different dopants in Si [17]

position in the lattice structure. The displacement creates an interstitial atom and a vacancy (i.e., a Frenkel pair), as shown in Fig. 3.6a. The displacement energy in silicon is approximately 15 eV [14]. At the beginning of the ion implantation process, lightweight ions have large energy where electronic collision are dominant and the energy loss due to nuclear collisions is smaller than the displacement energy

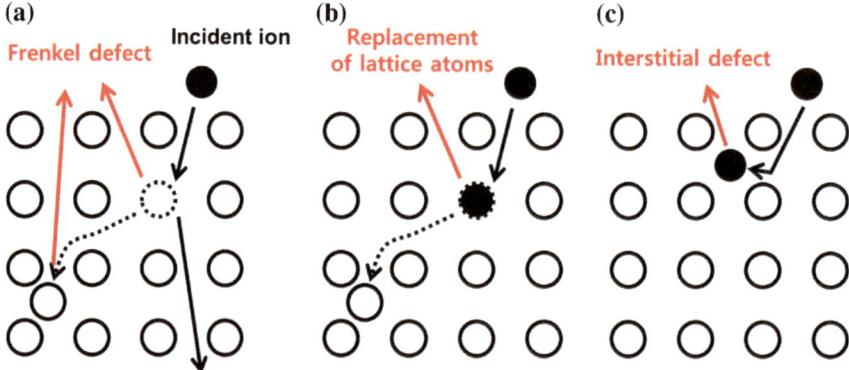

Fig. 3.6 Possible scenarios of collisions

(Fig. 3.5). As a result, they can deeply penetrate into the substrate without any implantation damage. When the lightweight ions have lost most of their energy (i.e., when they almost stop), the energy transferred through nuclear collisions is increased to such an extent as to displace target atoms. Therefore, lattice disorder induced by lightweight ions occurs far from the surface of the substrate, as shown in Fig. 3.4a. For heavy ions, substrate lattices are damaged by nuclear collisions close to the surface (Fig. 3.4b) because the energy loss due to the nuclear collision is much higher than the displacement energy (e.g., As in Fig. 3.5). On the other hand, if the incident ions have an energy level that is smaller than the displacement energy after a collision, there are two possible reasons: (1) replacement of lattice atoms (Fig. 3.6b), or (2) interstitial defects (Fig. 3.6c). For case (1), the incident ions land on the positions of target atoms and become interstitial defects.

Because the semiconductor substrate becomes amorphous by implantation disorders and interstitial defects, the mobility and lifetime of carriers in the semiconductor are degraded. Also, interstitial ions which do not have covalent bonds with silicon atoms cannot create mobile carriers. In other words, these impurities do not work as dopants at all. In order to repair the implantation damage and to activate the interstitial impurities, an annealing process follows the ion implantation process. However, not only does the annealing process restore the crystallinity of the semiconductor material and activate the impurities, it also diffuses those impurities into the substrate. Because the impurities randomly diffuse via a 'kick-out' mechanism, the annealing step also contributes to random dopant fluctuation [18–20]. However, the impact of the annealing process on the variability of dopant positions can be minimized through the use of advanced annealing techniques, such as rapid thermal annealing (RTA).

3.3 Characterization of Random Dopant Fluctuation (RDF)

Similar to the other process-induced random variations, the amount of RDF-induced performance variation can be simply and quantitatively estimated using the standard deviation. Assuming that the number of dopants in an individual device is determined by the Poisson distribution [11, 21], the standard deviation of the number of dopant atoms is easily calculated using the average number of dopants. For instance, as shown in Fig. 3.1, the average number of dopants in the channel region of transistors is 10^5 for the 10 μm technology node, and therefore, the corresponding standard deviation is $\sqrt{10^5}$. For the 100 nm technology node, the channel region of transistors has 100 dopants on average, and hence, the standard deviation of the number of dopants at the 100 nm technology node is $\sqrt{100}$. When compared to the relative standard deviation (RSD) (i.e., RSD = standard deviation/average), the RSD for the 100 nm technology node is higher than the RSD for the 10 μm technology node. This indicates that the distribution of the number of dopants is more dispersed from the average in the case of the 100 nm technology node. In other words, the random dopant fluctuation becomes worse as the physical channel length of the transistor is scaled down. However, this approach cannot completely describe the effects of random dopant fluctuation on transistor performance, such as threshold voltage, subthreshold slope, on-state current, and off-state leakage current, because the spatial positions of dopants are not fully considered. As mentioned in the previous section, the dopant atoms injected into a substrate undergo a series of collision events during the ion implantation process. Moreover, the dopant atoms diffuse into the substrate during annealing process. Because these collisions and diffusions randomly occur, the final distribution of dopant atoms varies from device to device, even if the total number of impurity atoms is identical.

Many researchers have investigated the effects of random dopant fluctuation on transistor performance using various approaches: (1) an analytical approach [22], (2) an atomistic process simulation using Kinetic Monte Carlo (KMC) simulation [23, 24], (3) a naïve approach [25], and (4) a full three-dimensional (3-D) TCAD simulation with randomized doping profiles [26–29]. In order to provide insight into random dopant fluctuation, two primary factors should be considered: (1) the number of dopant atoms, and (2) the positions of the given dopant atoms. In this chapter, we cover the Kinetic Monte Carlo simulation and the analytical approach, which are used to characterize process-induced random dopant fluctuation.

3.3.1 Kinetic Monte Carlo (KMC) Simulation

3.3.1.1 Kinetic Monte Carlo Process for Random Discrete Dopant Distribution

In order to simulate the threshold voltage variation induced by random dopant fluctuation, first of all, we need to determine the distribution for discrete dopants implanted into the substrate by the ion implantation process. There are some kinetic Monte Carlo simulation methods that are used to reasonably model the randomly distributed dopant atoms by ion implantation [30] and the annealing process [31]. First, in the simulation for the ion implantation process, the scattering angle and energy loss of impurity atoms are calculated using well-known conventional models [32]. Furthermore, small deflections due to both electronic collisions and nuclear collisions caused by the thermal vibration of lattice atoms are taken into account in the simulation [30]. Second, the diffusion behavior of dopant atoms in the annealing step is simulated using the random walk model [31]. Because the scattering angle of the trajectory and the diffusion length of the impurity atoms are randomly determined (i.e., even backward and/or diffused out of the simulation area), Kinetic Monte Carlo (KMC) simulation can implement a reasonable distribution for random discrete dopants with the statistics of (i) the positional fluctuation, and (ii) variations in the number of dopants (see Fig. 3.7).

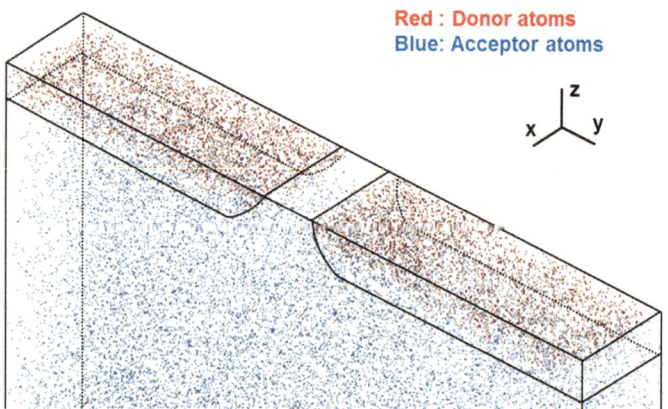

Fig. 3.7 A profile for random discrete dopants, generated by Kinetic Monte Carlo (KMC) simulation

3.3.1.2   Continuous Electric Potential for Calculating the Drift-Diffusion Equation

After creating the distribution profile for random discrete dopant atoms, we should convert it to the continuous electric potential in order to utilize the drift-diffusion mechanism. If we calculate the electric potential in the channel region using a point charge located at each ionized dopant atom, it results in non-physical singularities in the potential profile and charge density (see Fig. 3.8). Holes are easily trapped by negatively charged acceptors, leading to non-physical charge compensation, even in depletion regions. Sano et al. [33] and Ezaki et al. [34, 35] proposed a methodology to model the potential profile without the singularity problem. In Sano's model, the atomistic Coulomb potential is divided into long range and short range parts, and the long range part is only used to calculate the potential. The charge density is derived from the long range potential of a conduction electron as follows:

$$\rho(r) = \frac{qk_c^3 \{\sin(k_c r) - (k_c r)\cos(k_c r)\}}{2\pi^2 (k_c r)^3},\tag{3.2}$$

where k_c is the inverse screening length or the inverse of the Debye length [36]. The corresponding potential can be expressed as follows:

$$\phi(r) = \frac{ek_c}{2\pi^2 \varepsilon} \left\{ \frac{\mathrm{si}(k_c r)}{(k_c r)} - \sin(k_c r)^{\mathrm{r}} \right\},\tag{3.3}$$

where Si is the sine integral [34]. As shown in Fig. 3.9, the electric potential obtained from (3.3) approaches the bare potential when the distance is far from the original position of an ionized dopant atom (i.e., r = 0). However, the oscillation of the potential profile causes a large mismatch between Sano's model and the bare potential in the short range. The root cause of the oscillation in potential is the cosine term in (3.2). According to Ezaki et al. [34], the cosine term does not affect the total charge, so the second term in (3.2) and (3.3) can be removed. Thus, the charge density and the corresponding potential for each dopant atom can be

Fig. 3.8 Schematic diagram showing the effect of negatively charged acceptors on the electric potential. The long-range potential is used in simulation to remove unrealistic singularities in the potential and the charge density [33–35]

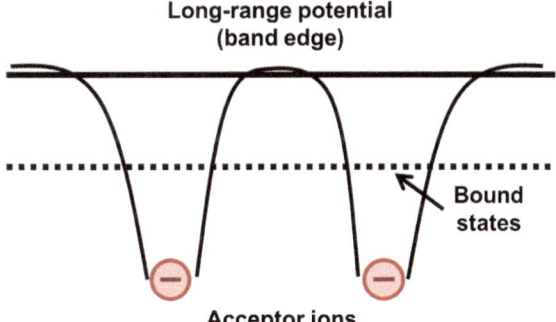

Fig. 3.9 Long-range part of
a charge density and
b potential, modeled by Ezaki
et al. [34] (*solid line*) and
Sano et al. [33] (*dashed line*)

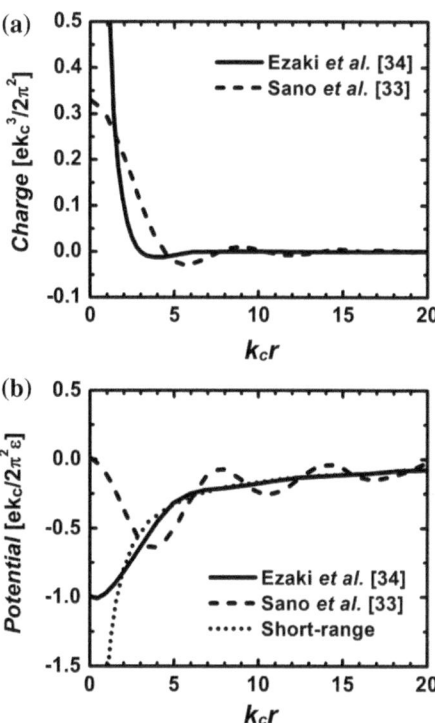

re-expressed as a function of the distance from the position of the dopant atom, as follows:

$$\rho(r) = \frac{ek_c^3\{\sin(k_c r) - (k_c r)\cos(k_c r)\}}{2\pi^2(k_c r)^3}, \tag{3.4}$$

$$\phi(r) = \frac{ek_v}{2\pi^2\varepsilon}\frac{si(k_v r)}{(k_c r)} \tag{3.5}$$

In order to more precisely estimate the threshold voltage variation induced by random dopant fluctuation, we have to carefully select the screening length. This is because the device fluctuation is extremely dependent on the screening length (Fig. 3.10). The screening length was determined to be 4 nm for doping concentrations higher than 10^{17} cm^{-3} [34].

3.3.1.3 Device Simulation Using Drift-Diffusion Transport

When the previously mentioned electric potential is set with a proper screening length, we can finally conduct Drift-Diffusion simulations to estimate the drain

Fig. 3.10 Impact of the screening length on the standard deviation of threshold voltage (*black square*), and threshold voltage (*red circle*) [34]

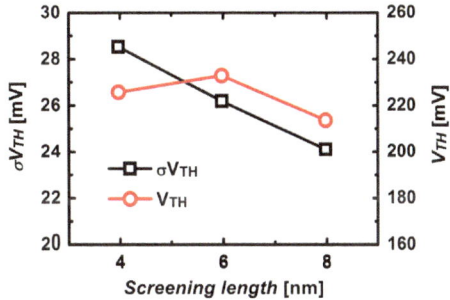

Table 3.1 Simulated standard deviation of threshold voltage and the number of dopants for n-type and p-type MOSFETs with different gate lengths [34]

MOSFET	n-type		p-type	
Gate length (nm)	40	70	35	60
σV_{TH} (mV)	59	28	40	27
σN_{dopant}	11	7	6	9

current of MOSFETs by solving Poisson's Equation (3.6) and the current-continuity equation (3.7), as follows:

$$\nabla \cdot (\varepsilon \nabla \psi) = q(n - p + N_A^- - N_D^+),\qquad(3.6)$$

$$\nabla \cdot J_n = 0,\qquad(3.7)$$

where ε is the dielectric permittivity; ϕ is the electric potential; n and p is electron and hole concentration, respectively; N_A^- and N_D^+ is acceptor and donor concentration, respectively; and J_n is electron current density. For example, Table 3.1 shows the number of dopants and the threshold voltage variation in n-/p-type MOSFETs with different gate lengths. An interesting fact shown in this table is that the variability in NMOS (gate length, $L_g = 40$ nm) is greater than that in PMOS (gate length, $L_g = 35$ nm). This is due to the fact that the fluctuation of channel dopant atoms in n-type MOSFETs is larger than that in p-type MOSFETs.

3.3.2 Analytical Model

Figure 3.11 shows the distribution of threshold voltages in 100,000 35 nm MOSFETs (gate length = 26 nm) and 140,000 13 nm MOSFETs (gate length = 8 nm). All the results in the figure are obtained from a three-dimensional (3-D) atomistic/statistical simulation [37]. It is well known that the doping concentration resulting from ion implantation follows the Gaussian distribution [15], and the

Fig. 3.11 Threshold voltage
distribution of MOSFETs
with different gate lengths:
a 35 nm device (gate length
of 26 nm) and **b** 13 nm
device (gate length of 8 nm)
[22]

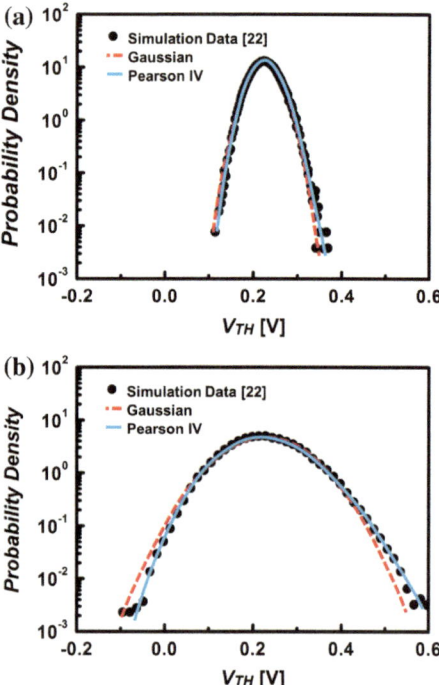

threshold voltage variation induced by random dopant fluctuation is also modeled
by the Gaussian distribution [27]. However, Fig. 3.11 clearly shows that the distribution of threshold voltages exactly matches the Pearson IV distribution because
of its asymmetry, [38] but is only a close match to the Gaussian distribution.
Although the Pearson IV distribution is useful for fitting the distribution of
threshold voltages, the relationship between this distribution and random dopant
fluctuation is unknown [22]. In this context, Reid et al. [22] suggested a new
analytical model that included various physical mechanisms of random dopant
fluctuation.

The asymmetry in the threshold voltage variation induced by random dopant
fluctuation originates from the fact that the number of dopant atoms in the channel
region is determined by Poisson's distribution [23, 39, 40]. In other words, because
the threshold voltage of the transistors is related to both the positions of the dopant
atoms and the number of dopant atoms, there is a mismatch between the various
distributions of threshold voltages (note that the Gaussian distribution only considers the position of dopant atoms). Similarly, when we model the random dopant
fluctuation using only the number of dopant atoms without considering their
positions, the results are not well matched to the actual distribution of threshold
voltages, as described in [41]. Before analyzing random dopant fluctuation using
both the numbers of dopant atoms and their positions, Reid et al., investigated the
impact of dopant atom positions on the threshold variation by fixing only the

(a)   **(b)**

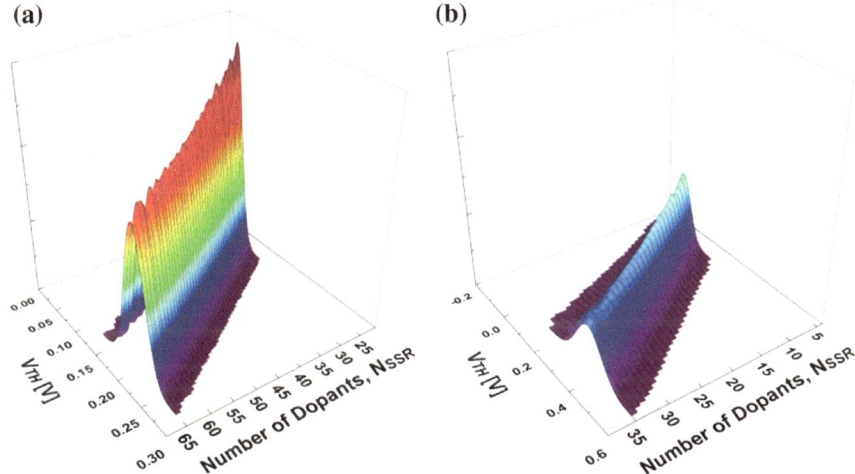

Fig. 3.12 Threshold voltage variation of **a** 35 nm devices and **b** 13 nm devices for the different number of dopants in the channel region (N_{SSR}) [22]

Fig. 3.13 Average and standard deviation of threshold voltage corresponding to Gaussian distribution in Fig. 3.12 [22]

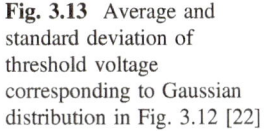

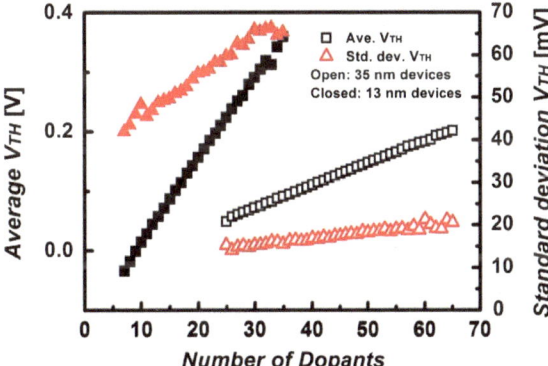

number of dopant atoms in the statistically significant region (N_{SSR}). The statistically significant region (SSR) indicates the region that dominates the statistical behavior of the device (i.e., the channel region). The distribution of threshold voltages for different N_{SSR} is plotted in Fig. 3.12a for 35 nm devices, and in Fig. 3.12b for 13 nm devices. Because the N_{SSR} is fixed, the results follow the Gaussian distribution, and hence, it is possible to estimate the mean and standard deviation. An interesting fact is that both the mean and the standard deviation of the threshold voltages are linearly proportional to N_{SSR} (Fig. 3.13).

In order to better describe the random dopant fluctuation, the number of dopant atoms should be taken into account. Because the probability that a lattice atom is replaced by a dopant atom is independent for all trials where lattice atoms are substituted with impurity atoms, we can say that a binominal distribution governs

Fig. 3.14 The estimated
results (using 3.9) are well
matched to simulated results

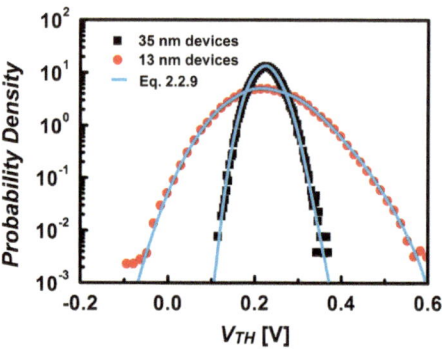

the number of dopant atoms. Moreover, for a limited case (i.e., infinite trials and a fixed mean), the Poisson distribution can be derived from the binomial distribution. In this case, the number of dopant atoms follows the Poisson distribution, since there are lots of trials of 10^5–10^6 and their mean is fixed to average the doping concentration. Therefore, the complete description for the distribution of threshold voltages can be obtained by multiplying the Poisson distribution [i.e., $f(N_{SSR}, \overline{N_{SSR}})$] with the Gaussian distribution [i.e., $g(V_{TH}, \mu_{N_{SSR}}, \sigma_{N_{SSR}})$] for a given N_{SSR} (3.8) and then summing (3.8) for every N_{SSR}, as follows:

$$P(V_{TH}|N_{SSR}) = f(N_{SSR}, \overline{N_{SSR}}) \cdot g(V_{TH}, \mu_{N_{SSR}}, \sigma_{N_{SSR}}), \tag{3.8}$$

$$P(V_{TH}) = \sum_{i=N_{SSRmin}}^{N_{SSRmax}} f(i, \overline{N_{SSR}}) \cdot g(V_{TH}, \mu_{N_{SSR}}, \sigma_{N_{SSR}}), \tag{3.9}$$

where f is the probability mass function of the Poisson distribution; $\overline{N_{SSR}}$ is the average number of dopant atoms in SSR; g is the probability density function of the Gaussian distribution; and $\mu_{N_{SSR}}$ and $\sigma_{N_{SSR}}$ are the expected value and the standard deviation of the threshold voltage for a given N_{SSR}, respectively.

The distribution of threshold voltages obtained by (3.9) is exactly matched to the simulation results (as shown in Fig. 3.14). Moreover, it should be noted that the expected value and the standard deviation of threshold voltages are monotonically increasing with increasing N_{SSR} (Fig. 3.13). Therefore, the expected value and the standard deviation for various N_{SSR} values can be easily extrapolated without a lot of simulation. The distribution of threshold voltages calculated by extrapolation is well matched to the simulation results [22]. Therefore, one can make the conclusion that the threshold voltage variation induced by random dopant fluctuation is simply and easily estimated by simulation for only two given N_{SSR}s (i.e., in order to generate the required minimum number of points for the linear/monotonic extrapolation).

References

1. Moore GE (1975) Progress in digital integrated electronics. In: IEDM technical digest, pp 11–13
2. Yau LD (1974) A simple theory to predict the threshold voltage for short-channel IGFET's. Solid-State Electron 17(10):1059–1063
3. Yan R-H, Ourmazd A, Lee KF (1992) Scaling the Si MOSFET: from bulk to SOI to bulk. IEEE Trans Electron Devices 39(7):1704–1710
4. Hanafi HI, Noble WP, Bass RS, Varahramyan K, Lii Y, Dally AJ (1993) A model for anomalous short-channel behavior in submicron MOSFET's. IEEE Electron Device Lett 14 (12):575–577
5. Sadana D, Acovic A, Shahidi G, Hanafi H, Warren A, Grutzrnacher D, Cardone F, Sun J, Davari B (1992) Enhanced short-channel effects in NMOSFET's due to boron redistribution induced by arsenic source and drain implant. In: IEDM technical digest, pp 849–852
6. Gwoziecki R, Skotnicki T, Bouillon P, Gentil P (1999) Optimization of V_{TH} roll-off in MOSFET's with advanced channel architecture—retrograde doping and pockets. IEEE Trans Electron Devices 46(7):1551–1561
7. Troutman RR (1979) VLSI limitations from drain-induced barrier lowering. IEEE Trans Electron Devices 26(4):461–469
8. Fjeldly TA, Shur M (1993) Threshold voltage modeling and the subthreshold regime of operation of short-channel MOSFETs. IEEE Trans Electron Devices 40(1):137–145
9. Chamberlain SG, Ramanan S (1986) Drain-induced barrier-lowering analysis in VSLI MOSFET devices using two-dimensional numerical simulations. IEEE Trans Electron Devices 33(11):1745–1753
10. Dennard RH, Gaensslen FH, Yu H-N, Rideout VL, Bassous E, LeBlanc AR (1974) Design of ion-implanted MOSFET's with very small physical dimensions. IEEE J Solid-State Circ 9 (5):256–268
11. Nam H, Lee GS, Lee H, Park IJ, Shin C (2014) Analysis of random variations and variation-robust advanced device structures. J Semicond Technol Sci 14(1):8–22
12. Zhao W, Cao Y (2006) New generation of predictive technology model for sub-45 nm early design exploration. IEEE Trans. Electron Device 53(11):2816–2823
13. Kuhn K, Kenyon C, Kornfeld A, Liu M, Maheshwari A, Shih W-K, Sivakumar S, Taylor G, VanDerVoorn P, Zawadzki K (2008) Managing process variation in Intel's 45 nm CMOS technology. Intel Technol. J. 12(2):93–109
14. Plummer JD, Deal MD, Griffin PB (2000) Silicon VLSI technology: fundamentals, practice and modeling. Prentice Hall, New Jersey
15. Jaeger RC (2002) Introduction to microelectronic fabrication, 2nd edn. Prentice Hall, New Jersey
16. Brice DK (1975) Recoil contribution to ion implantation energy deposition distribution. J Appl Phys 46(8):3385–3394
17. Sze SM (2008) Semiconductor devices: physics and technology, 2nd edn. Wiley, New Jersey
18. Nichols CS, Van de Walle CG, Pantelides ST (1989) Mechanisms of dopant impurity diffusion in silicon. Phys Rev B 40(8–15):5484–5497
19. Nichols CS, Van de Walle CG, Pantelides ST (1989) Mechanisms of equilibrium and nonequilibrium diffusion of dopants in silicon. Phys Rev Lett 62(9–27):1049–1052
20. Cowern NEB, van de Walle GFA, Gravesteijn DJ, Vriezema CJ (1991) Experiments on atomic-scale mechanisms of diffusion. Phys Rev Lett 67(2–8):212–215
21. Bukhori MF (2011) Simulation of charge-trapping in nano-scale MOSFETs in the presence of random-dopants-induced variability. PhD thesis, University of Glasgow
22. Reid D, Millar C, Roy G, Roy S, Asenov A (2009) Analysis of threshold voltage distribution due to random dopants: A 100,000-sample 3-D simulation study. IEEE Trans Electron Devices 56(10):2255–2263

23. Wong H.-SP, Taur Y (1993) Three-dimensional "atomistic" simulation of discrete random dopant distribution effects in sub-0.1 μm MOSFETs. In: IEDM technical digest, pp 705–708
24. Frank DJ, Taur Y, Ieong N, Wong H-SP (1999) Monte Carlo modeling of threshold variation due to dopant fluctuations, In: Symposium on VLSI technical digest, pp 169–170
25. Li Y, Yu S-M, Hwang J-R, Yang F-L (2008) Discrete dopant fluctuations in 20-nm/15-nm-gate planar CMOS. IEEE Trans Electron Devices 55(6):1449–1455
26. Asenov A, Slavcheva G, Brown AR, Davies JH, Saini S (2001) Increase in the random dopant induced threshold fluctuations and lowering in sub-100 nm MOSFETs due to quantum effects: a 3-D density-gradient simulation study. IEEE Trans Electron Devices 48(4):722–729
27. Asenov A (1998) Random dopant induced threshold voltage lowering and fluctuations in sub-0.1 m MOSFET's: A 3-D "atomistic" simulation study. IEEE Trans Electron Devices 45 (12):2505–2513
28. Asenov A, Brown AR, Davies JH, Kaya S, Slavcheva G (2003) Simulation of intrinsic parameter fluctuations in decananometer and nanometer-scale MOSFETs. IEEE Trans Electron Devices 50(9):1837–1851
29. Brown AR, Asenov A, Watling JR Intrinsic fluctuations in sub 10-nm double-gate MOSFETs introduced by discreteness of charge and matter. IEEE Trans. Nanotechnol 1(4):195–200
30. Hane M, Fukuma M (1990) Ion implantation model considering crystal structure effects. IEEE Trans Electron Devices 37(9):1959–1963
31. Hane M, Ikezawa T, Takeuchi K, Gilmert GH (2001) Monte Carlo impurity diffusion simulation considering charged species for low thermal budget sub-50 nm CMOS process modeling. In: IEDM Technical Digest, pp 38.4.1–38.4.4
32. Ziegler JF, Biersack JP (1985) The stopping and range of ions in matter. Springer, New York
33. Sano N, Matsuzawa K, Mukai M, Nakayama N (2000) Role of long-range and short-range coulomb potentials in threshold characteristics under discrete dopants in sub-0.1 μm Si-MOSFETs. In: IEDM technical digest, pp 275–278
34. Ezaki T, Ikezawa T, Hane M (2002) Investigation of realistic dopant fluctuation induced device characteristics variation for sub-100 nm CMOS by using atomistic 3D process/device simulator. In: Proceedings of IEEE IEDM, pp 311–314
35. Ezaki T, Ikezawa T, Notsu A, Tanaka K, Hane M (2002) 3D MOSFET simulation considering long-range Coulomb potential effects for analyzing statistical dopant-induced fluctuations associated with atomistic process simulator. In: Proceedings of SISPAD, pp 91–94
36. Shin C, Sun X, Liu T-JK (2009) Study of random-dopant-fluctuation (RDF) effects for the trigate bulk MOSFET. IEEE Trans Electron Devices 56(7):1538–1542
37. Roy G, Brown AR, Adamu-Lema F, Roy S, Asenov A (2006) Simulation study of individual and combined sources of intrinsic parameter fluctuations in conventional nano-MOSFETs. IEEE Trans Electron Devices 53(12):3063–3070
38. Heinrich J (2004) A guide to the Pearson type IV distribution. University of Pennsylvania, Philadelphia, PA, Technical Report
39. Millar C, Reid D, Roy G, Roy S, Asenov A (2008) Accurate statistical description of random dopant-induced threshold voltage variability. IEEE Electron Device Lett 29(8):946–948
40. Kovac U, Reid D, Millar C, Roy G, Roy S, Asenov A (2008) Statistical simulation of random dopant induced threshold voltage fluctuations for 35 nm channel length MOSFET. Microelectron Reliab 48(8/9):1572–1575
41. Toriyama S, Sano N (2003) Probability distribution functions of threshold voltage fluctuations due to random impurities in deca–nano MOSFETs. Phys E 19(1/2):44–47

Chapter 4
Work Function Variation (WFV)

4.1 Introduction

The channel length of metal oxide semiconductor field effect transistors (MOSFETs) has been continuously and successfully scaled down over the past few decades, at the pace described by Moore's Law. However, aggressively scaled channel lengths have caused MOSFETs to become more vulnerable to short channel effects (SCEs) and process-induced random variations, as discussed in previous sections. In an attempt to alleviate these undesirable effects, the gate-to-channel capacitance has been increased by using a thinner gate oxide layer in order to enhance gate control over the channel potential. However, in extremely scaled MOSFETs, gate oxide scaling is no longer viable because it requires a sub-1 nm gate oxide thickness that corresponds to only a few layers of SiO_2. But, sub-1 nm SiO_2 is too thin to fabricate easily, and it causes a significant amount of gate leakage current. In sub-60 nm technology nodes, it has been predicted that the gate leakage current will surpass the subthreshold leakage current unless the sub-1 nm thick SiO_2 layer can be replaced with an electrically thin but physically thick insulation layer [1, 2]. As a result, high-k (HK) dielectric material has been adopted to limit ever-increasing gate leakage currents and worsened short channel effects because the HK material can act as both an electrically thin and a physically thick dielectric layer and provide better gate-to-channel control over the channel potential in MOSFETs [3–6]. In 2007, high-k/metal-gate (HK/MG) technology was introduced for the first time in the 45 nm complementary metal oxide semiconductor (CMOS) technology node [7].

Although HK/MG technology has been adopted for state-of-the-art MOSFETs, it is inevitably accompanied by threshold voltage (V_{TH}) variation induced by the work function variation (WFV) in the metal gate. When HK dielectric material is integrated into the MOS structure with a polysilicon gate, MOSFETs suffer from Fermi-level pinning and phonon scattering because of poor compatibility between the HK dielectric material and the polysilicon gate. Oxygen vacancies at the interface between the HK dielectric material and the polysilicon gate lead to

© Springer Science+Business Media Dordrecht 2016
C. Shin, *Variation-Aware Advanced CMOS Devices and SRAM*,
Springer Series in Advanced Microelectronics 56,
DOI 10.1007/978-94-017-7597-7_4

Fig. 4.1 Three-dimensional isometric view of a planar bulk MOSFET, showing the WFV of TiN metal gate. It is generated by the TCAD simulation tool

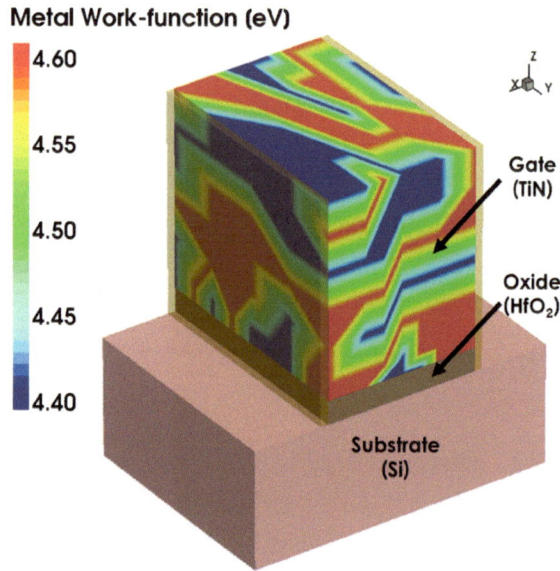

Fermi-level pinning, which degrades performance in MOSFETs (i.e., equivalently, higher threshold voltage) [8, 9]. Phonon scattering happens when optical phonon vibration interferes with the electrons' movement in the channel region, and causes their mobility to be degraded [10]. In order to overcome these technical issues, metals such as TiN and TaN are used for the gate electrode in place of the polysilicon material [11–13]. However, the use of a metal gate creates a new random V_{TH} variation because the local work function of the metal gate depends on the grain orientation of each grain (Fig. 4.1) [14, 15], and a metal gate consists of many grains which are randomly sized with different grain orientations as a result of the atomic layer deposition (ALD) process used to create the metal gate [16]. In other words, the value of the metal work function depends on the grain orientation of each metal grain. However, because the size and orientation of the metal grains are randomly determined, the overall work function for the metal gate (or the average work function for the metal gate) is not identical across all the MOSFETs in an integrated circuit (IC). In other words, device-to-device work function variation exists, and the threshold voltages of MOS devices in an IC are affected by the work function variation of the metal gate [17–19].

In a way that is similar to the other random variation sources discussed in previous sections, the impact of WFV on V_{TH} variation can be neglected when the gate electrode area is large enough to include lots of grains (i.e., the averaging effect). However, in sub-32 nm technology nodes, the size of the gate electrode area is quite close to the average grain size (i.e., average grain size of TiN ~ 22 nm), and the amount of V_{TH} variation induced by WFV is no longer negligible [20]. Hence, compared to the other random variation sources described, the total V_{TH} variation is dominated by WFV.

4.2   The Physical Origins of WFV

It is commonly known that the work function is defined as the minimum energy required to move an electron from solid state material to vacuum. In other words, the work function can be interpreted as the energy difference between the Fermi Energy level and the vacuum energy level. However, the work function cannot be simply defined as the energy difference between the Fermi Energy level and the vacuum energy level of a solid state material because the surface dipole potential increases the work function. The dependence of the surface dipole potential on the surface charge density leads to variations in the work function for different crystal orientations.

4.2.1   Characteristics of Metal Grains

The atoms that make up metal material are placed alongside neighboring atoms and arranged in order. These atoms form crystal structures in which unit cells are periodically positioned (e.g., simple cubic (SC), body-centered cubic (BCC), face-centered cubic (FCC), diamond structure, and zinc blende structure, etc.). The metal used in the gate electrodes of MOSFETs consists of small crystal grains together with defects and disorientations. These small crystal grains have various sizes and grain orientations relative to each other. During the thermal treatment process for the gate stack in MOSFETs (i.e., the annealing process), small grains tend to combine with neighboring grains and grow up into a larger grain (see Fig. 4.2) [21].

Depending on the deposition and annealing process conditions, the grain size can grow up to 5–20 nm in typical IC fabrication steps [17]. In order words, the grain size of metals in the gate stack is comparable to the physical gate length of

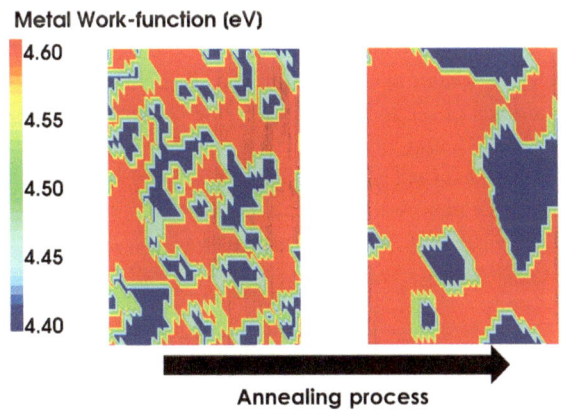

Fig. 4.2 Annealing process where small grains in a metal gate are combined into a large grain

Metal Work-function (eV)

4.60

4.55

4.50

4.45

4.40

Annealing process

Fig. 4.3 XRD plot of an
arbitrary material. By
comparing the intensities in
the XRD plot, we can
calculate the probability that
each grain is formed

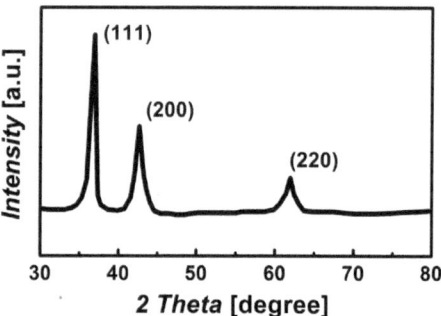

MOSFETs in sub-20 nm semiconductor technology, and this indicates that only a
few grains exist at the bottom-side surface of the gate electrode.

During the time that the metal grains are combining with adjacent grains, the
metal grains are simultaneously recrystallized in certain orientations. Although
metal grains tend to crystalize into more stable orientations at very high temperature, they are likely to enlarge themselves in stable and other crystal orientations at
the lower temperature ranges used in IC fabrication steps. By investigating the
results of X-ray diffraction (XRD) (Fig. 4.3), we can experimentally measure the
distributions of grains (i.e., grain size distribution and grain orientation distribution). A strong XRD peak indicates that most metal grains are crystalized in the
corresponding orientation. On the other hand, a weak XRD peak means that the
number of metal grains with the corresponding orientation is relatively small.
Therefore, the probability of finding a particular grain orientation can be obtained
using the XRD method. Because the work function and the orientation of grains are
in one-to-one correspondence, the probability of grain orientation is critical in
characterizing the work function variation (WFV).

4.2.2   Dependence of the Metal Work Function on the Grain Orientation

The orientation of a metal grain is defined as the vector that is perpendicular to the
surface of the metal grain (i.e., the vector that is normal to the metal grain plane). In
an FCC unit cell, the cubic structure can be cut by various planes, and each
cross-sectional plane is characterized by its own grain orientation. For instance, the
surface planes of metal grains can be represented by planes with grain orientation of
$\langle x, y, z \rangle$ (e.g., $\langle 100 \rangle$, $\langle 110 \rangle$, and $\langle 111 \rangle$, as shown in Fig. 4.4).

Now, we can come up with a concept of surface charge density that indicates the
number of metal atoms per unit area of the surface plane. Regardless of grain
orientation, the effective number of atoms (when only taking into account the part
of the atom inside the unit cell) is identical, but it is different for the surface area.
Thus, it is straightforward to deduce that the surface charge density simply relies on

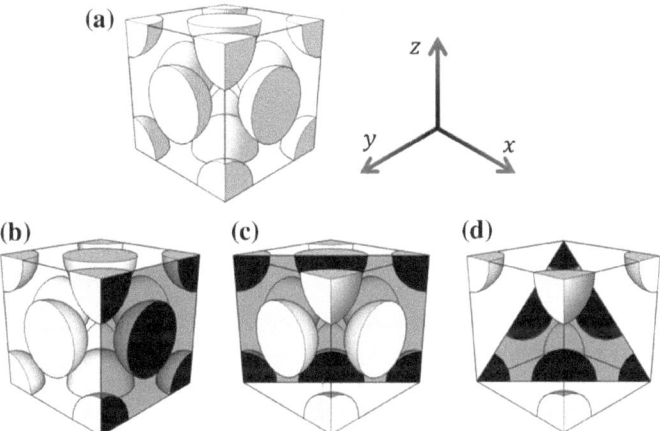

Fig. 4.4 **a** Arrangement of atoms in an FCC unit cell. The effective number of atoms (*black mark*) which are on the **b** $\langle 100 \rangle$, **c** $\langle 110 \rangle$, and **d** $\langle 111 \rangle$ planes is identical regardless of the direction of the surface plane

Table 4.1 Surface charge density of an FCC unit cell with various surface plane directions (herein, the side length of the FCC unit cell is a)

Plane direction	Effective number of atoms	Plane area	Surface charge density
$\langle 100 \rangle$	2	a^2	$2/a^2$
$\langle 110 \rangle$	2	$\sqrt{2}a^2$	$1.41/a^2$
$\langle 111 \rangle$	2	$\sqrt{3}a^2/2$	$2.31/a^2$

the grain orientation. Estimated surface charge densities with different orientations are summarized in Table 4.1.

The dependency of the metal work function on grain orientation originates from its surface dipole potential, which was introduced in [22]. Lang and Kohn [23] explained the surface dipole potential using electron density. According to the jellium model, the electron density distribution spreads out over the metal surface to which the electron is associated (see Fig. 4.5).

In other words, some electrons can be observed outside the metal surface, and thereby, can create negative charge. Since metals are electrically neutral, in general, the total number of positive and negative charges should be balanced. However, because electrons are distributed beyond the metal surface, the balance between the number of electrons and the number of metal ions is broken by the existence of electrons beyond the metal surface. As a consequence, a positive charge appears inside the metal surface or close to the surface. An interesting fact is that electron

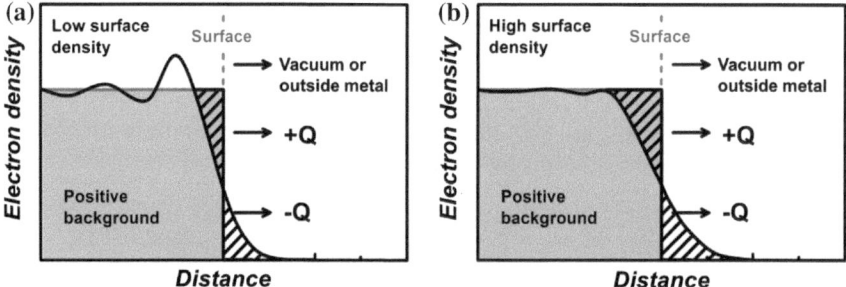

Fig. 4.5 Electron distribution at the surface of metal with **a** low surface charge density and **b** high surface charge density. It should be noted that a stronger dipole is formed when the surface charge density is higher

emission from the Fermi Energy level to the vacuum energy level can be blocked by dipoles, which originate from the two opposite charges at the surface. Therefore, the work function is increased as the dipoles get stronger. In other words, the dipoles get stronger when the surface charge density of the metal grains increases because more charged electrons are outside of the metal. Herein, it is noteworthy that the surface charge density of metal grains with $\langle 111 \rangle$ orientation is highest, followed by the surface charge density with orientations of $\langle 100 \rangle$ and $\langle 110 \rangle$. This was verified by experimental measurement, as shown in [24].

4.2.3 Impact of WFV on V_{TH} Variation

In order to investigate the impact of WFV on threshold voltage (V_{TH}) variation in MOSFETs, the V_{TH} of MOSFETs should be analytically expressed in an equation. The V_{TH} of MOSFETs is typically expressed by

$$V_{TH} = V_{FB} + 2\phi_F \pm \frac{\varepsilon_{si}}{\varepsilon_{ox}} t_{ox} \sqrt{\frac{4qN_B}{\varepsilon_{si}\varepsilon_o}} (\pm\phi_F) \begin{array}{l} (+)\text{for n–channel devices} \\ (-)\text{for p–channel devices} \\ N_B = N_A \text{ or } N_D \end{array} , \quad (4.2.1)$$

where V_{FB} is flat band voltage; ϕ_F is bulk potential; t_{ox} is oxide thickness; and ε_{si}, ε_{ox} and ε_o are the permittivity of silicon, oxide and vacuum, respectively. V_{FB} is the voltage required to eliminate band bending in the MOS structure (i.e., it is defined as the energy difference that is between the work function values of metal and semiconductor (or silicon)). As shown in (4.2.1), the work function variation (WFV) directly and linearly (at least, monotonically) affects V_{TH} variation.

4.3 Characterization of WFV

4.3.1 Statistical Analysis

As mentioned in the previous section, the physical gate length of semiconductor technology nodes (i.e., 20–32 nm technology and below) is quite comparable to the average size of metal grains, which grow up to 5–20 nm in CMOS fabrication processes. Thus, the number of grains existing in metal gates decreases in the order of 10–100, causing WFV-induced V_{TH} variations to become worse. To address this technical issue, many researchers have tried to establish an analytical model that quantitatively and simply characterizes the WFV. However, because (i) the grain orientation is randomly determined in CMOS fabrication processes, and (ii) it is difficult to control the grain orientation, the work function distribution can only be modeled as a stochastic or probabilistic distribution. To put it simply, the WFV can be characterized and quantitatively estimated in a unit of standard deviation.

Dadgour et al. [25] statistically analyzed the WFV by modeling the probability distribution of the work function of metal gates. In order to estimate the WFV of metal gates, work function values of grains with different orientations, and the corresponding probabilities, and grain sizes, are necessary. The different work function values of each grain, which are represented by Φ_i (where i = 1, 2, …, n), can be obtained by capacitance-voltage (C-V) measurements [26]. It should be noted that C-V measurements are conducted on the samples, which are composed of a single metal grain (not multiple metal grains), in order to obtain the work function value of a particular/single grain. In contrast to CMOS processes, the maximum temperature in the metal deposition steps when fabricating the sample should not be restricted by the thermal budget, if possible, thereby allowing the grains to grow to cover the whole area of the sample. Thus, C-V measurements for single large grains provide the work function value of each grain orientation. From the XRD plot, the probability (i.e., P_i, where i = 1, 2, …, n) of a certain work function value (Φ_i) and grain size (G) can be extracted. The intensity of XRD peaks for various orientations can be used to calculate the probability of finding each grain orientation in the metal gate material, and the width of the XRD peak shows the average grain size of the corresponding grain orientation. Note that all grains with different orientations are assumed to have an identical grain size in order to have a low complexity and closed-form analytical solution. Assuming square-shaped grains, the total number of grains (N) that exist in the metal gate area (L × W) can be determined as (L/G) × (W/G).

The total work function of the metal gate composed of grains can be calculated as follows:

$$\Phi_M = \left(\frac{X_1}{N}\right)\Phi_1 + \left(\frac{X_2}{N}\right)\Phi_2 + \cdots + \left(\frac{X_n}{N}\right)\Phi_n, \qquad (4.3.2)$$

where X_1, X_2, ..., X_n are random variables, which represent the number of grains with work function values of Φ_1, Φ_2, ..., Φ_n, respectively. Notice that X_1/N indicates the probability (P_1) of the work function value (Φ_1) when the total number of grains is large. Thus, if we know the occurrence probability of each work function, the mean and standard deviation for Φ_M can be easily calculated. If a metal gate is composed of two grain orientations only, the total work function can be written as follows:

$$\Phi_M = \left(\frac{X_1}{N}\right)\Phi_1 + \left(\frac{X_2}{N}\right)\Phi_2 \qquad (4.3.3)$$

Given the total number of grains, if the distribution of the random variable X_1 is known, the distribution of the other random variable X_2 is also known because the metal gate in this example is composed of only two grain orientations (i.e., the total number of grains ($X_1 + X_2$) is equal to N). Since the probability of work function value Φ_1 is equal for all grains (i.e., independent of each other), the number of grains with work function value Φ_1 (X_1) follows the binomial distribution. The probability of "$X_1 = k$" is given by the probability mass function as follows:

$$f_{X_1}(k) = \binom{N}{k}P_1^k(1 - P)^{N-k}, \quad \text{where} \quad \binom{N}{k} = \frac{N!}{k!(N - k)!} \qquad (4.3.4)$$

For example, a TiN metal gate consists of $\langle 100 \rangle$ and $\langle 111 \rangle$ grains with work functions of 4.6 and 4.4 eV, and probability of 60 and 40 %, respectively. In the case of a metal gate with four grains (N = 3), possible values of Φ_M and the corresponding probability of X_1 (herein, the number of grains with a work function of 4.6 eV) can be calculated using (4.3.3) and (4.3.4), respectively, as shown in Table 4.2.

Figure 4.6 shows the distribution functions for the WF of metal gates with different numbers of grains. The distribution of work functions with higher numbers of grains can also be obtained by using (4.3.3) and (4.3.4) for all possible situations/cases. It is noteworthy that the distribution of the work function is similar to the Gaussian distribution when the number of grains increases, as predicted by the Central Limit Theorem [27].

Table 4.2 Possible compositions of a metal gate that has three grains with two orientations, and the corresponding probabilities and metal work functions

(X_1, X_2)	Probability	Work function (eV)
(0, 3)	$\frac{3!}{0!(3-0)!} \times 0.6^0 \times 0.4^{(3-0)} = 0.064$	$\left(\frac{0}{3}\right)4.6 + \left(\frac{3}{3}\right)4.4 = 4.4$
(1, 2)	$\frac{3!}{1!(3-1)!} \times 0.6^1 \times 0.4^{(3-1)} = 0.288$	$\left(\frac{1}{3}\right)4.6 + \left(\frac{2}{3}\right)4.4 = 4.46$
(2, 1)	$\frac{3!}{2!(3-2)!} \times 0.6^2 \times 0.4^{(3-2)} = 0.432$	$\left(\frac{2}{3}\right)4.6 + \left(\frac{1}{3}\right)4.4 = 4.53$
(3, 0)	$\frac{3!}{3!(3-3)!} \times 0.6^3 \times 0.4^{(3-3)} = 0.216$	$\left(\frac{3}{3}\right)4.6 + \left(\frac{0}{3}\right)4.4 = 4.6$

Fig. 4.6 The distribution functions for the WF of metal gates with various numbers of grains

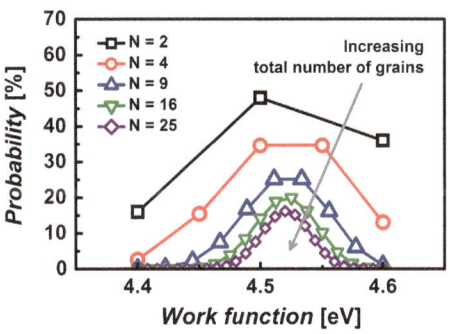

In order to obtain the probability distribution of the work functions for three or more grain orientations, one can apply the generalized form of the binomial distribution, a.k.a., the 'multinomial distribution'. However, in contrast to the case of two grain orientations, the random variables (i.e., X_1, X_2, and X_3) are independent of each other. In other words, the distribution of X_1 cannot determine the distribution of the other variables. Thus, the previous method is not enough to fully characterize the probability distribution of individual random variables (i.e., X_1, X_2, and X_3). Hence, the distribution of the work function (Φ_M) cannot be obtained in a closed form, unfortunately. However, if the number of grains is large enough (≈ 10–15) to warrant application of the Central Limit Theorem [27], the probability distribution of work functions follows the Gaussian distribution, and therefore, is fully characterized using expected value and variance (i.e., mean and standard deviation). The generalized expected value and variance for the work function with more than three grain orientations is provided as follows:

$$E(\Phi_M) = \sum_{i=1}^{r} P_i \phi_i \tag{4.3.5}$$

$$\text{var}(\Phi_M) = \frac{1}{N} \left[\sum_{i=1}^{r} P_i \phi_i^2 - \left(\sum_{i=1}^{r} P_i \phi_i^2 \right)^2 \right]. \tag{4.3.6}$$

The detailed calculation steps are specified in [25]. The model has also been verified by Monte Carlo simulation. As shown in the equation for variance, aggressively-scaled devices (i.e., smaller N) suffer from severe WFV.

4.3.2 Ratio of Average Grain Size to Gate Area (RGG)

Since the scaling limit of the grain size in widely used metal gate materials (e.g., TiN, TaN, WN, and MoN) exists to some degree, the physical upper boundary of the

WFV of those metal gate materials needs to be determined. Then, others can suggest/provide the corresponding device design guidelines. Although Dadgour et al. [25] proposed the analytical model for WFV, it is self-limited by the assumption that all the metal grains have identical size, leading to a Gaussian distribution of work functions. In the annealing process, small grains tend to merge together to form large grains of inconsistent size. For this reason, the Rayleigh distribution is more realistic and suitable for modeling WFV-induced random V_{TH} fluctuation than is the Gaussian distribution [28]. A three-dimensional (3-D) simulation using a more realistic granular grain with random grain size and polygon shape was carried out in order to statistically study V_{TH} variation in scaled decananometer High-k/Metal-Gate (HK/MG) MOSFETs [18]. However, the method requires an unacceptable amount of time to be spent on computing the WFV-induced V_{TH} variation. In this context, for the purpose of simplicity and to accurately estimate the WFV-induced V_{TH} variation, a new concept [i.e., RGG, or ratio of average grain size to gate area, equal to {average grain size in nm} times {(channel length × channel width)$^{-1/2}$ in nm^{-1}}] was introduced by Nam and Shin [19].

Figure 4.7 shows the threshold voltage (V_{TH}) distribution for MOSFETs with different average grain sizes and TiN metal gate areas. In the case of 35 × 35 nm gate areas with average grain sizes of 5 nm, most V_{TH} values for devices are positioned in the center of the distribution. However, as the grain size increases, the distribution spreads out, and V_{TH} values are concentrated and/or peaked at the

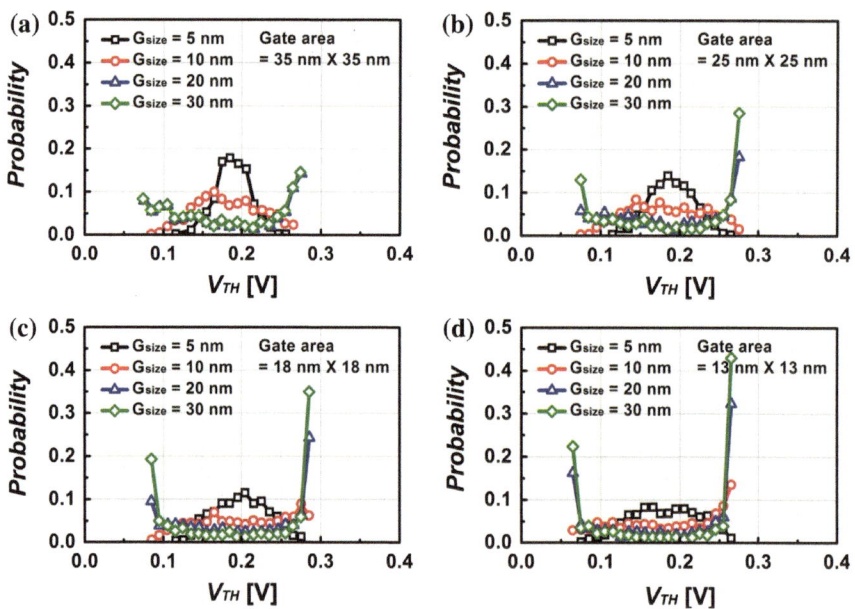

Fig. 4.7 Threshold voltage distribution for MOSFETs which has **a** 35 nm × 35 nm, **b** 25 nm × 25 nm, **c** 18 nm × 18 nm, and **d** 13 nm × 13 nm metal gate [18]. G_{size} is average grain size

bounded values. Notice that the voltage difference between the maximum and minimum value of V_{TH} is 0.2 V because the difference between the work functions for two grain orientations of TiN is 0.2 eV. This tendency is the same for other cases and materials. On the other hand, for a given average grain size, the variability of V_{TH} increases because the number of grains in the gate area decreases with decreasing gate area. Therefore, one can conclude that the distribution of WFV depends on both the average grain size and the gate area, and therefore, the concept of 'RGG' can be simply used to quantitatively estimate V_{TH} variability in CMOS HK/MG devices.

In order to verify the validity of the RGG concept in estimating WFV, TiN metal gates with two different gate areas (i.e., 146.7 and 28.7 nm^2), and two average grain sizes (i.e., 22 and 4.3 nm), but the same value of RGG (herein, RGG = 0.15), are generated using experimental data, such as the grain orientations, their corresponding work function and probability values [29, 30], and average grain sizes [17]. Based on the Rayleigh distribution, the grains fill the gate area, and therefore, all of the devices cannot have the identical number of grains in their gate areas. The total number of grains for 100,000 samples, each with uniquely randomized grains, is shown in Fig. 4.8.

It is noteworthy that the distributions of total grains are nearly equal to each other in spite of different gate areas and different average grain sizes. Therefore, it is expected that, if the RGG value of the devices is given, we can estimate the amount of WFV regardless of device size because the WFV is highly dependent on the number of grains (4.3.6).

Finally, the amount of WFV can be obtained by calculating the work function for each device. The calculated standard deviation of WFV (i.e., σWFV) for

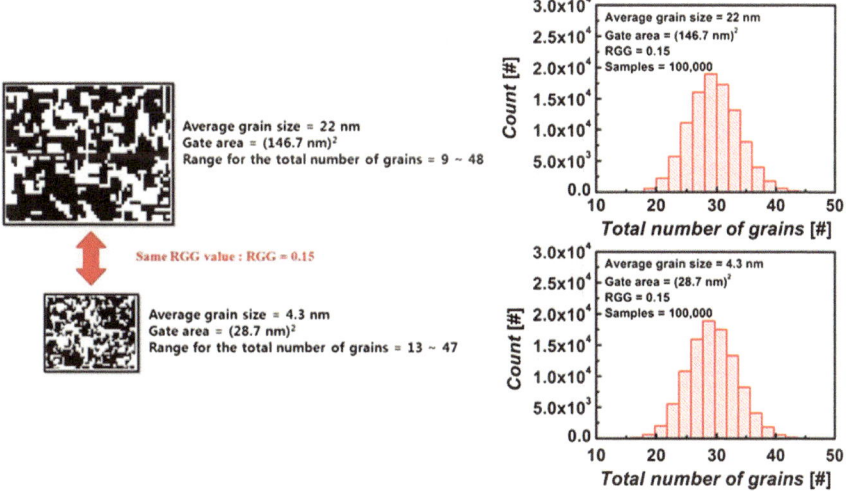

Fig. 4.8 For given RGG value (but with different gate area and average grain sizes), the total number of grains in metal gate is shown

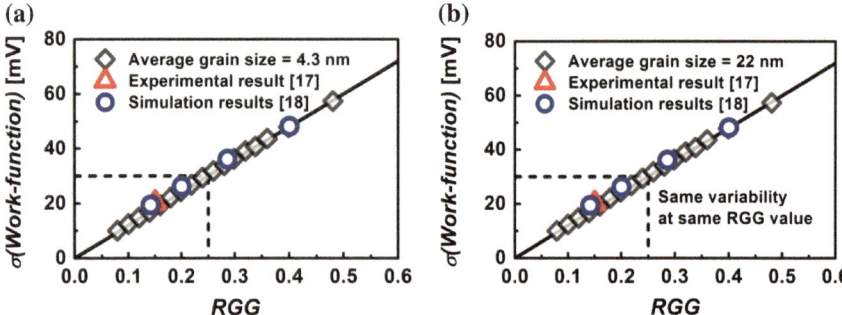

Fig. 4.9 The amount of WFV is plotted against the RGG for different average grain sizes:
a 4.3 nm and **b** 22 nm [19]

different average grain sizes is plotted against RGG in Fig. 4.9. By varying the gate area for a given grain size, TiN metal gates with various RGGs are generated. All the gates used in computing σWFV are also composed of grains following the Rayleigh distribution. It should be noted that the different average grain size is used to calculate the WFV, but the same RGG, and then σ(WFV) are obtained. Based on the fact that the RGG plot is exactly matched to the experimental data [17] and simulation results [18], it confirms that the estimated WFV based on the Rayleigh distribution for the grain sizes and the RGG concept is verified and valid. Therefore, once the gate area and average grain size are chosen, the amount of WFV in a given semiconductor technology can be easily detected by simply tracing the RGG plot without time-consuming simulations or measurements. Furthermore, when both the gate material and variation corners [i.e., σ(WFV)] are provided, the minimal gate area can be easily determined. Moreover, the grain size to meet both the minimal gate area (limited by photolithography technique) and the variation corners (tightened by the targeting yield) can also be easily extracted.

Figure 4.10 shows the σ(WFV) versus the RGG plot for a wider range. It provides a standard for determining: (1) to what extent the WFV should be controlled and (2) to what extent the grain size should be minimized in upcoming advanced semiconductor technology. Note that, when the gate area becomes smaller than

Fig. 4.10 Standard deviation of WFV is plotted against the RGG with a wider range

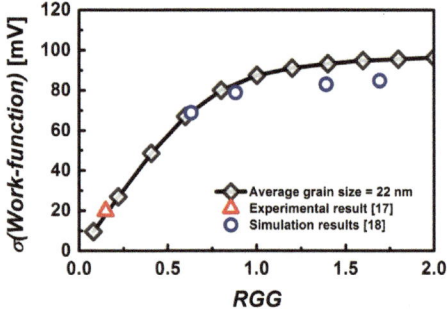

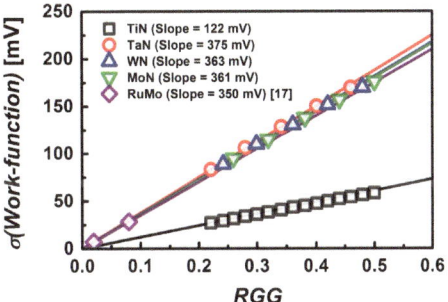

Fig. 4.11 RGG plot for various metal materials. Slopes of the RGG plot for TaN, WN, MoN, and RuMo are comparable

average grain size (i.e., RGG > 1), σ(WFV) begins to saturate at a particular value; if the average grain size is larger than the gate area, RGG would overestimate the WFV-induced V_{TH} variation because the gate area is filled with only a few large grains [19].

The σ(WFV) versus RGG plot with different metal materials is shown in Fig. 4.11. Even though the slope of the RGG plot varies depending on what kind of materials are used, it is still independent of the RGG value (i.e., the slopes for TiN, TaN, WN, MoN, and RuMo are 122, 375, 363, 361, and 350 mV, respectively). This is because of the number of grain orientations that metals can have, and their corresponding probability and work function difference from material to material. Therefore, the WFV-induced threshold voltage variation can be easily and quantitatively estimated using the RGG plot, and ultimately, the optimal gate area or average grain size can be obtained, which satisfies the required variability specification for digital logic operations.

References

1. De V, Borkar S (1999) Technology and design challenges for low power and high performance. In: Proceedings of the 1999 international symposium on low power electronics and design, pp 163–168
2. Hicks J, Bergstrom D, Hattendorf M, Jopling J, Maiz J, Pae S, Prasad C, Wiedemer J (2008) 45 nm transistor reliability. Intel Technol J 12(2):131–144
3. Pidin S, Morisaki Y, Sugita Y, Aiyama T, Irino K, Nakamura T, Sugii T (2002) Low standby power CMOS with HfO₂ gate oxide for 100-nm generation. In: Symposium on VLSI technology of digest, pp 28–29
4. Morisaki Y, Aoyama T, Sugita Y, Irino K, Sugii T, Nakamura T (2002) Ultra-thin (T_{eff}^{inv} = 1.7 nm) poly-Si-gated SiN/HfO₂/SiON high-k stack dielectrics with high thermal stability (1050 °C). In: Proceedings of IEEE IEDM, pp 861–864
5. Kang L, Onishi K, Jeon Y, Lee BH, Kang C, Qi W-J, Nieh R, Gopalan S, Choi R, Lee JC (2000) MOSFET devices with polysilicon on single-layer HfO₂ high-k dielectrics. In: IEDM technical digest, pp 35–38
6. Lee SJ, Luan HF, Bai WP, Lee CH, Jeon TS, Senzaki Y, Roberts D, Kwong DL (2000) High quality ultra thin CVD HfO₂ gate stack with poly-Si gate electrode. In: IEDM technical digest, pp 31–34

7. Mistry K, Allen C, Auth C, Beattie B, Bergstrom D, Bost M, Brazier M, Buehler M, Cappellani A, Chau R, Choi C-H, Ding G, Fischer K, Ghani T, Grover R, Han W, Hanken D, Hattendorf M, He J, Hicks J, Huessner R, Ingerly D, Jain P, James R, Jong L, Joshi S, Kenyon C, Kuhn K, Lee K, Liu H, Maiz J, McIntyre B, Moon P, Neirynck J, Pae S, Parker C, Parsons D, Prasad C, Pipes L, Prince M, Ranade P, Reynolds T, Sandford J, Shifren L, Sebastian J, Seiple J, Simon D, Sivakumar S, Smith P, Thomas C, Troeger T, Vandervoorn P, Williams S, Zawadzki K (2007) A 45 nm logic technology with high-k+metal gate transistors, strained silicon, 9 Cu interconnect layers, 193 nm dry patterning, and 100% Pb-free packaging. In: Proceedings of IEEE IEDM, pp 247–250

8. Takeuchi H, Wong HY, Ha D, Liu T-JK (2004) Impact of oxygen vacancies on high-κ gate stack engineering. In: IEDM technical digest, pp 829–832

9. Hobbs CC, Fonseca LRC, Knizhnik A, Dhandapani V, Samavedam SB, Taylor WJ, Grant JM, Dip LG, Triyoso DH, Hegde RI, Gilmer DC, Garcia R, Roan D, Lovejoy ML, Rai RS, Hebert EA, Tseng H-H, Anderson SGH, White BE, Tobin PJ (2004) Fermi-level pinning at the polysilicon/metal oxide interface: Part I. IEEE Trans Electron Devices 51(6):971–977

10. Gusev EP, Narayanan V, Frank MM (2006) Advanced high-k dielectric stacks with poly-Si and metal gates: recent progress and current challenges. IBM J Res Develop 50(4/5):387–410

11. Datta S, Dewey G, Doczy M, Doyle BS, Jin B, Kavalieros J, Kotlyar R, Metz M, Zelick N, Chau R (2003) High mobility Si/SiGe strained channel MOS transistors with HfO₂/TiN gate stack. In: IEDM technical digest, pp 28.1.1–28.1.4

12. Wang X, Liu J, Zhu F, Yamada N, Kwong D-L (2004) A simple approach to fabrication of high-quality HfSiON gate dielectrics with improved nMOSFET performances. IEEE Trans Electron Devices 51(11):1798–1804

13. Chau R, Datta S, Doczy M, Doyle B, Kavalieros J, Metz M (2004) High-k/metal–gate stack and its MOSFET characteristics. IEEE Electron Devices Lett 25(6):408–410

14. Dadgour HF, Endo K, De V, Banerjee K (2008) Modeling and analysis of grain-orientation effects in emerging metal-gate devices and implications for SRAM reliability. In: Proceedings of IEE IEDM, pp 1–4

15. Dadgour HF, Endo K, De VK, Banerjee K (2010) Grain-orientation induced work function variation in nanoscale metal-gate transistors—Part II: implications for process, device, and circuit design. IEEE Trans Electron Devices 57(10):2515–2525

16. Frye A, Galyon GT, Palmer L (2007) Crystallographic texture and whiskers in electrodeposited thin films. IEEE Trans Electron Packag Manuf 30(1):2–10

17. Ohmori K, Matsuki T, Ishikawa D, Morooka T, Aminaka T, Sugita Y, Chikyow T, Shiraishi K, Nara Y, Yamada K (2008) Impact of additional factors in threshold voltage variability of metal/high-k gate stacks and its reduction by controlling crystalline structure and grain size in the metal gates. In: Proceedings of IEEE IEDM, pp 1–4

18. Wang X, Brown AR, Idris N, Markov S, Roy G, Asenov A (2011) Statistical threshold-voltage variability in scaled decananometer bulk HKMG MOSFETs: a full-scale 3-D simulation scaling study. IEEE Trans Electron Devices 58(8):2293–2301

19. Nam H, Shin C (2013) Study of high-k/metal-gate work-function variation using Rayleigh distribution. IEEE Electron Devices Lett 34(4):532–534

20. Asenov A, Slavcheva G, Brown AR, Davies JH, Saini S (2001) Increase in the random dopant induced threshold fluctuations and lowering in sub-100 nm MOSFETs due to quantum effects: a 3-D density-gradient simulation study. IEEE Trans Electron Devices 48(4):722–729

21. Grovenor CRM, Hentzell HTG, Smith DA (1984) The development of grain structure during growth of metallic films. Acta Mater 32(5):773–781

22. Smoluchowski R (1941) Anisotropy of the electronic work function of metals. Phys Rev 60 (9):661–674

23. Lang N, Kohn W (1970) Theory of metal surfaces: charge density and surface energy. Phys Rev B 1(12):4555–4568

24. Gaillard N, Mariolle D, Bertin F, Gros-Jean M, Proust M, Bsiesy A, Bajolet A, Chhun S, Djebbouri M (2006) Characterization of electrical and crystallographic properties of metal

layers at deca-nanometer scale using Kelvin probe force microscope. Microelectron Eng 83 (11–12):2169–2174

25. Dadgour HF, Endo K, De VK, Banerjee K (2010) Grain-orientation induced work function variation in nanoscale metal-gate transistors—Part I: modeling, analysis, and experimental validation. IEEE Trans Electron Devices 57(10):2504–2514

26. Buiu O, Hall S, Engstrom O, Raeissi B, Lemme M, Hurley PK, Cherkaoui K (2007) Extracting the relative dielectric constant for 'high-κ layers' from CV measurements—errors and error propagation. Microelectron Reliab 47(4–5):678–681

27. Feller W (1971) An introduction to probability theory and its applications, 3rd edn. Wiley, New York

28. Nam H, Shin C (2013) Comparative study in work-function variation: Gaussian vs. Rayleigh distribution for grain size. IEICE Electron Express 10(9):20130109

29. Yagishita A, Saito T, Nakajima K, Inumiya S, Matsuo K, Shibata T, Tsunashima Y, Suguro K, Arikado T (2001) Improvement of threshold voltage deviation in damascene metal gate transistors. IEEE Trans Electron Devices 48(8):1604–1611

30. Hussain MM, Quevedo-Lopez MA, Alshareef HN, Wen HC, Larison D, Gnade B, El-Bouanani M (2006) Thermal annealing effects on physical properties of a representative high-k/metal film stack. Semicond Sci Technol 21(10):1437–1440

Part II
Variation-Aware Advanced CMOS Devices

Chapter 5
Tri-Gate MOSFET

5.1 Introduction

As a result of the continuous and successful advancements in the field of complementary metal oxide semiconductor (CMOS) technology, the half pitch for a memory cell had reached down to 32 nm by 2009 [1]. However, scaling down the gate length of the metal oxide semiconductor field effect transistor (MOSFET) results in the generation of process-induced random variations [2–4] and severe short-channel effects [5–12] such as threshold voltage roll-off, exponentially increasing off-state leakage current, degraded subthreshold slope (SS), and drain-induced barrier lowering (DIBL). Thus, it is extremely challenging to reduce the minimum feature size of a planar bulk MOSFET by a factor of 0.7 for the next generation CMOS technology, while retaining good electrostatic integrity. This is the reason behind the adoption of tri-gate MOSFETs with three-dimensional (3-D) device structures in the 22-nm technology node [13]. The non-planar structure (in which, the self-aligned gate straddles the fin-shaped narrow silicon channel) can strengthen the gate control over the channel potential. In addition, because the fin-shaped silicon channel is narrow enough to be fully depleted, the tri-gate MOSFET shows better immunity against short-channel effects without heavily doping the channel region. In other words, there is no need for heavy doping of the channel to alleviate short-channel effects. The superior gate controllability not only improves the electrostatic integrity of the tri-gate MOSFET but also provides higher tolerance to process-induced random variations, when compared to the planar bulk MOSFET. For example, in a conventional planar bulk MOSFET, the threshold voltage variation induced by the random dopant fluctuation (RDF) is the main technical issue related to the modern CMOS technologies [14], and it affects the static random access memory (SRAM) yield critically [15]. However, because the channel region of the tri-gate MOSFET is lightly doped (or almost undoped), the use of a tri-gate MOSFET can suppress the effects of RDF on the performance of

© Springer Science+Business Media Dordrecht 2016
C. Shin, *Variation-Aware Advanced CMOS Devices and SRAM*,
Springer Series in Advanced Microelectronics 56,
DOI 10.1007/978-94-017-7597-7_5

CMOS logic devices or SRAM bit cells. Therefore, tri-gate MOSFETs support sustainable CMOS scaling in accordance with the Moore's law.

Although the impact of RDF on the threshold voltage variations is significantly reduced in the tri-gate MOSFET, variations resulting from the lightly doped channel region (with channel doping concentration below 10^{17} cm^{-3}) and the heavily doped source/drain regions still exist. These variations should be estimated accurately for the ultimate miniaturization of CMOS devices. Moreover, the unique device structure of the tri-gate MOSFET changes the characteristics of the LER (Line Edge Roughness) and WFV (Work-Function Variation): for example, the narrow fin-shaped silicon channel generates new random variation sources during the fabrication process, such as fin LER and sidewall LER, contrary to the gate-edge roughness, which is present in the conventional LER [16]. The amount of WFV in the tri-gate MOSFET is different from that in the conventional planar bulk MOSFET, despite the identical RGG values (RGG is the ratio of average grain size to gate area), because the metal gate wraps around the channel rather than being on the surface [17]. Furthermore, the WFV-induced variations vary with the shape of the current flow (e.g., single bulky current flow vs. isolated surface current flow) [18]. In this chapter, the performance variations induced by the RDF, LER, and WFV in tri-gate MOSFETs will be addressed to provide an insight on how to overcome these variations.

5.2   RDF in Tri-Gate MOSFET

Figure 5.1 shows the simulated I_{DS}-V_{GS} curves for a planar bulk MOSFET and a tri-gate bulk MOSFET [19]. To generate a randomly distributed atomistic doping profile in the body and source/drain regions, the methodology suggested in [20] is used. The I_{DS}-V_{GS} curves for both the planar bulk MOSFET and the tri-gate bulk MOSFET, with the generated atomistic doping profiles, are obtained using the Sentaurus^{TM} technology computer-aided design (TCAD; which is one of the widely used commercial device simulators). Note that the channel region in the tri-gate MOSFET does not have to be heavily doped in order to prevent short-channel effects. As shown in Fig. 5.1, the variation in the off-state leakage current can be reduced significantly by introducing the tri-gate device structure. Furthermore, the amount of threshold voltage variation (i.e., standard deviation of threshold voltage, σV_{TH}) is decreased by 30 % in the tri-gate bulk MOSFET, compared to the planar bulk MOSFET. This can be explained by the equation for σV_{TH} given below [21]:

$$\sigma V_{TH} = \frac{\overline{Q_B}}{2 C_{OX} \frac{W_{eff}}{W_{stripe}} \sqrt{W_{stripe} L_{eff} \overline{Q_B}/q}} \tag{5.2.1}$$

where, W_{eff} and W_{stripe} are the effective and layout widths of the channel, respectively, L_{eff} is the distance between two points where the source/drain doping

Fig. 5.1 Simulated I_{DS}
versus V_{GS} curves for **a** planar
bulk MOSFET and **b** tri-gate
bulk MOSFET [19]

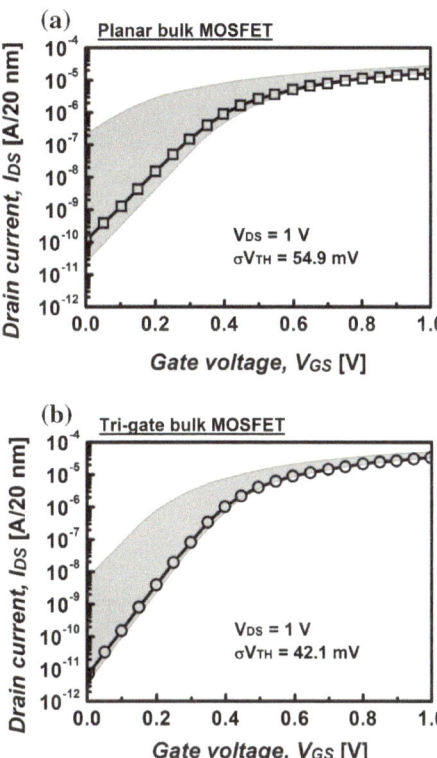

concentration is equal to 2×10^{19} cm^{-3}, and $\overline{Q_B}$ is the depletion charge per unit
layout area (the derivation of 5.1.1 is described in [19, 21]). Although the depletion
charge is increased slightly because of the 3-D channel geometry, a wider effective
channel leads to lesser threshold-voltage variations.

The contribution of body RDF (source/drain RDF) to the total RDF can be
estimated using the device simulation based on only the body RDF (source/drain
RDF). Figure 5.2 shows the standard deviations of the threshold voltages for planar

Fig. 5.2 Standard deviation
of threshold voltage in planar
bulk MOSFET and tri-gate
bulk MOSFET, under various
RDF conditions [19]

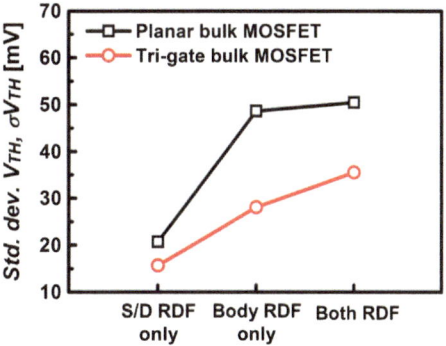

and tri-gate MOSFETs under various RDF conditions. In both the device structures, the impact of body RDF is larger than that of source/drain RDF because the doping profile in the channel region plays a more critical role in determining the threshold voltage of the MOSFET. Contrary to the planar MOSFET, the total RDF in the tri-gate MOSFET is significantly higher than the body RDF. Considering the lighter doping concentration in the channel region of the tri-gate MOSFET, the source/drain doping profile is as important as the body doping profile (it will be covered later in this section).

The tri-gate device structure is beneficial for the threshold voltage adjustment because it provides a new way to shift the threshold voltage. In the conventional planar bulk MOSFET, the channel doping concentration is increased to increase the threshold voltage. However, as shown in Fig. 5.3, this approach inevitably increases the RDF in the channel region, because the depletion charge in 5.1.1 is increased by increasing the channel doping concentration. For the tri-gate MOSFET, the threshold voltage can be adjusted by tuning the peak depth of the retrograde body doping (herein, the peak depth of the retrograde body doping profile is identical to the physical height of the fin, in order to achieve the best performance [22]). Because the increase in the depletion charge is relatively reduced by varying the depth, we can increase the threshold voltage with a smaller increase in the RDF, when compared to the conventional threshold voltage adjustment method.

In terms of the variations in DIBL and SS, the tri-gate MOSFET is more robust to RDF than the planar MOSFET (see Fig. 5.4, [23]). To investigate the effect of RDF on MOSFETs, a channel region is created with randomly distributed dopants: (1) Dopants are randomly located in a large cube of volume 96 nm^3 with an average doping concentration of 1.5×10^{18} cm^{-3}. (2) Sub-cubes of volume 16 nm^3 are extracted from the large cube and inserted into the channel region of the devices-under-simulation. It is noteworthy that the nominal devices are carefully designed to have the same channel lengths and widths of 16 nm, and threshold voltages of 250 mV. The variations in a tri-gate bulk MOSFET with an aspect ratio (i.e., fin height/fin width) of 1 is decreased by 55.4 % in DIBL and 20.7 % in SS,

Fig. 5.3 Effect of threshold voltage adjustment by: (*square*) varying the doping concentration or (*circle*) modifying the fin height, on the RDF-induced threshold voltage variations [19]

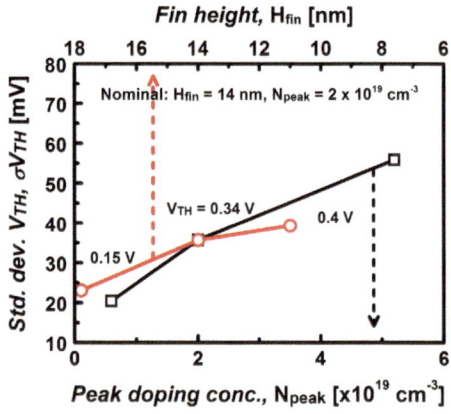

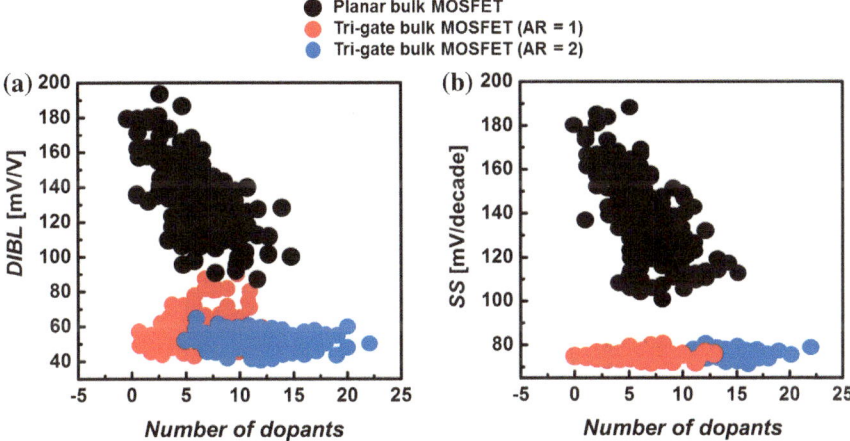

Fig. 5.4 RDF-induced **a** DIBL and **b** SS variations for planar bulk MOSFET and tri-gate bulk MOSFET with two different aspect ratios (i.e., aspect ratio of 1 and 2) [23]

compared to the planar bulk MOSFET. The tri-gate bulk MOSFET with an aspect ratio of 2 shows more improvement: 68.7 % in DIBL and 30.1 % in SS. The basis for the implementation of the variation-robust tri-gate bulk MOSFET is (i) superior gate controllability and (ii) higher gate oxide capacitance of the tri-gate bulk MOSFET (compared the planar bulk MOSFET). However, the position of dopants in the channel region highly affects the DIBL variation. Figure 5.5 shows the channel potential of two different cases, in which the DIBL values are different even though the same number of dopants is used. When the dopants exist near the channel surface, the DIBL increases because the channel potential fluctuates significantly with changes in the external drain voltage. However, if the dopants are

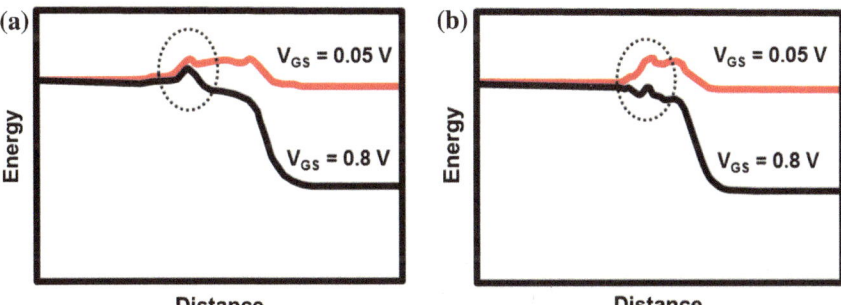

Fig. 5.5 Channel potential of two different cases: most dopants are located **a** away from and **b** near the channel surface. Note that the two cases have the same number of dopants, but different DIBL values [23]

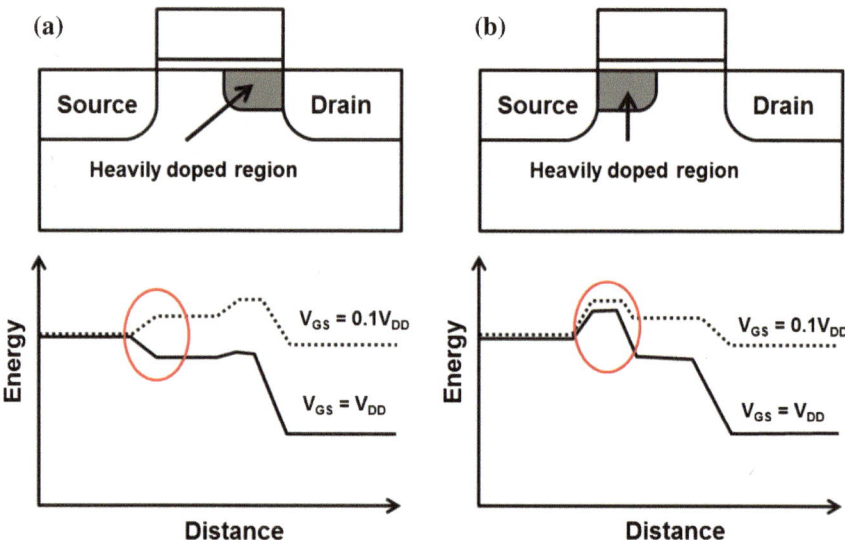

Fig. 5.6 Cross-sectional view of MOSFET with asymmetric doping profiles and the corresponding channel potential plots

away from the channel surface, the fluctuations in the channel potential are relatively smaller, and thereby, the tri-gate MOSFET shows a lower DIBL. Moreover, the non-uniformity of the source-to-drain doping profile can also cause DIBL variations: a high doping concentration at the drain-to-channel region degrades the DIBL, whereas a high doping concentration at the source-to-channel region improves the DIBL [24] (Fig. 5.6).

As discussed above, the RDF-induced threshold voltage variations are dramatically decreased in the tri-gate MOSFET (compared to the planar MOSFET) because of the lightly doped or undoped channel region. However, because the tri-gate MOSFET is still exposed to the RDF in the source/drain regions, the device structure of the tri-gate MOSFET should be further optimized. As shown in Fig. 5.7, a large number of dopants are randomly distributed in the source/drain gradient regions. If the source/drain dopants spread deeper into the channel region (i.e., more gradual source/drain doping gradient), the effective channel length is shortened and the RDF-induced threshold voltage variation is increased (see 5.2.1). Therefore, a steeper source/drain doping gradient is necessary to suppress the RDF-induced threshold voltage variation (see Fig. 5.8). Moreover, the steeper source/drain doping gradient is beneficial in terms of (i) lower variations in on/off currents and (ii) lesser short-channel effects [25].

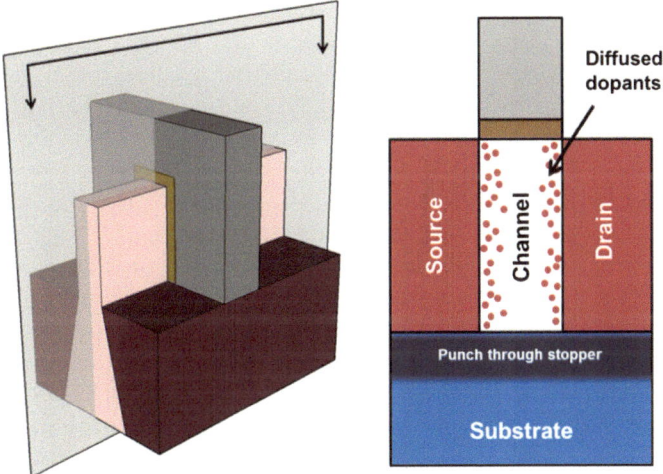

Fig. 5.7 Dopants in source/drain regions are diffused into the channel region. Note that these diffused dopants cause RDF, especially in the source/drain gradient region

Fig. 5.8 Standard deviation of threshold voltage as a function of source/drain doping abruptness [25]

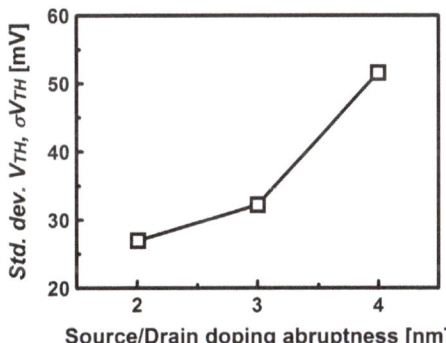

5.3 LER in Tri-Gate MOSFET

In order to enhance the gate-to-channel capacitive coupling, as well as to suppress the short-channel effects in the tri-gate MOSFET, a taller and narrower fin-shaped silicon channel region is necessary. However, a narrower channel region produces a new random-variation source named as the fin line-edge-roughness (LER) because the impact of the fin roughness on the channel width is increased as the channel becomes narrow. In addition, the intrinsic limit of the etching technique creates a roughness along the channel height direction (which is called sidewall LER). These new random variation sources in the tri-gate device structure can directly deform the physical geometry of the fin-shaped channel region, whereas the conventional gate LER changes the gate length only. Because the device performance metrics and short-channel effects of the tri-gate MOSFET are closely related to the physical geometry of the channel region, the impact of LER on the tri-gate MOSFET becomes severe. Therefore, the amount of LER-induced threshold voltage variation is comparable to the WFV-induced threshold voltage variation [26]. Moreover, the LER is not scaled down in the advanced technology nodes [2]. In this context, the LER and the tri-gate MOSFET design with deformed geometries were studied in various researches [16, 27–32].

The nominal tri-gate bulk MOSFET is illustrated in Fig. 5.9. First, a trapezoidal channel region (instead of the rectangular channel region) is fabricated using the 22-nm tri-gate technology, and then the rectangular channel region is fabricated using the 14-nm 2nd-generation tri-gate technology [33]. Because of corner effects in the fin body [13], the impact of the fin angle (θ) on the LER for a given top-width of the channel should be considered. To implement the effects of LER on the tri-gate bulk MOSFET, random edge profiles following the Gaussian distribution with a standard deviation of 1 nm are generated and mapped onto the nominal tri-gate bulk device. Figure 5.10 shows the four possible LERs in a tri-gate MOSFET: fin LER with resist patterning and spacer patterning, sidewall LER, and gate LER. The fin LER with resist patterning indicates the fluctuation of the channel width generated by conventional lithography techniques using photoresist. On the contrary, the fin LER with spacer patterning refers to the LER occurring because of the self-aligned double patterning using a dummy spacer. Thus, the LER profiles at the two side-edges of the channel (or fin) are uncorrelated when using resist patterning, but closely correlated when using spacer patterning [34].

Fig. 5.9 Three-dimensional (3-D) bird's-eye view of tri-gate bulk MOSFET

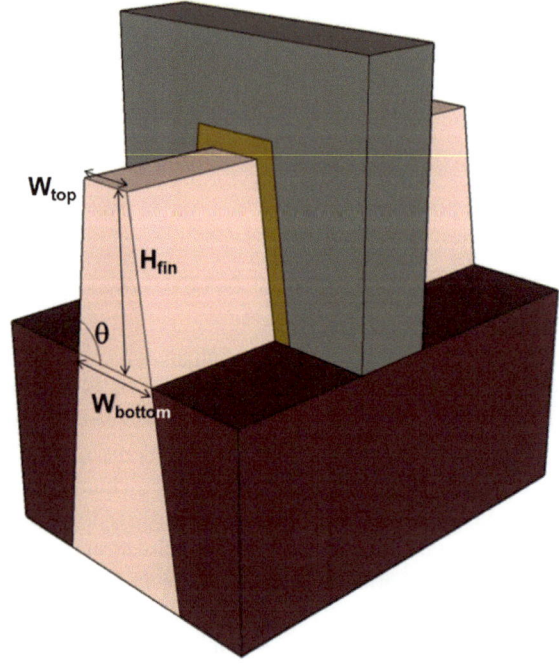

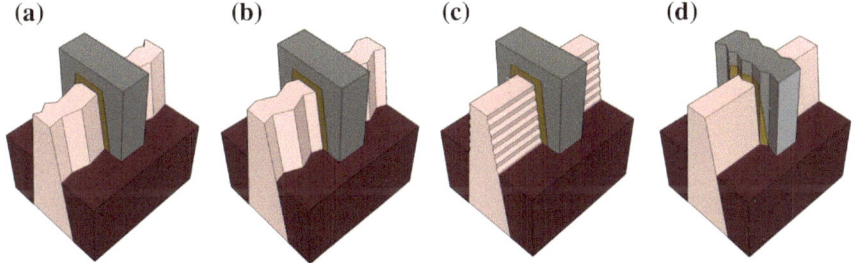

Fig. 5.10 3-D bird's-eye views of tri-gate bulk MOSFET with **a** fin LER induced by resist patterning, **b** fin LER induced by spacer patterning, **c** sidewall LER, and **d** gate LER

Figure 5.11 shows the variations in the DIBL value (i.e., ΔDIBL/DIBL) against the channel-width variation for various fin angles. The DIBL variations tend to increase as the channel width increases because of weak gate controllability over the channel. The sensitivity of the DIBL to channel-width variations becomes worse with steeper fin angles. This originates from a larger variation in the inversion charge, which can be controlled by the gate [16].

The LER-induced threshold voltage variations for various fin angles are shown in Fig. 5.12. Because the channel width varies greatly in the channel-length direction, the carriers suffer from surface roughness scattering and the fin LER with resist patterning induces larger threshold voltage variations [16]. Similarly, for the sidewall LER, because the channel width fluctuates along the channel-height direction, the LER-induced threshold voltage variations increase because of the surface roughness scattering. However, in case of the gate LER, the channel width remains the same along channel width/height directions. Although the gate length of the transistor fluctuates, the LER-induced threshold voltage variations are small because of the superior gate-to-channel coupling. The fin LER with spacer patterning shows the least amount of threshold voltage variations among the four LER sources. The reason behind this least LER-induced threshold voltage variation is the uniform channel width. In order words, the channel width remains the same in the channel-length

Fig. 5.11 DIBL variation as a function of width variation for various fin angles [16]

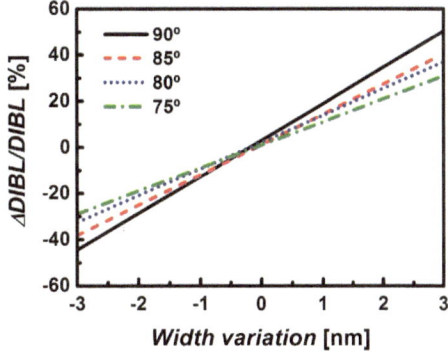

Fig. 5.12 Standard deviation
of threshold voltage as a
function of fin angle for four
different LER sources [16]

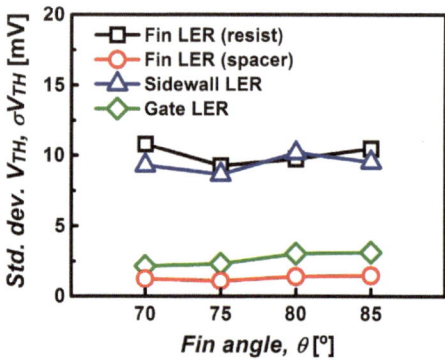

direction because of the strong correlation between the two side-edges of the channel,
and thereby, the surface roughness scattering and the LER-induced threshold voltage
variation can be alleviated. It is noteworthy that the LER-induced threshold voltage
variation is almost independent of the fin angle (see Fig. 5.12).

Figure 5.13a–d shows the on/off-state current variations in the four LER cases,
for various fin angles. Although the on-state current decreases with increasing the

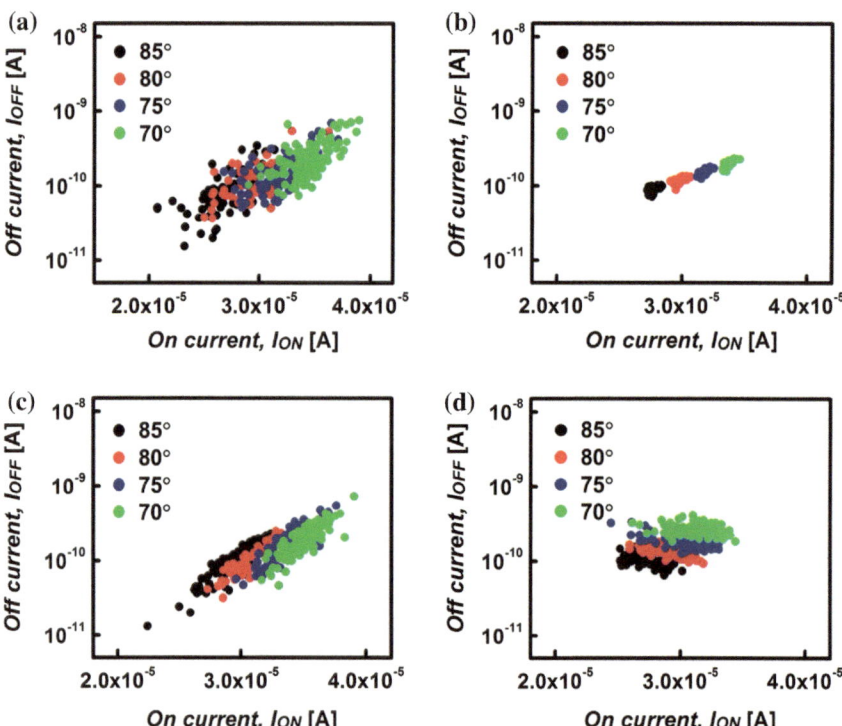

Fig. 5.13 On and off current variations by **a** fin LER induced by resist patterning, **b** fin LER
induced by spacer patterning, **c** sidewall LER, and **d** gate LER, for various fin angles [16]

fin angle because of the large effective width, the on/off-current ratio is increased because the off-state current decreases more rapidly because of strong gate controllability. Similar to the threshold voltage variations, the amount of on/off-state current variations induced by the fin LER with resist patterning is comparable to that induced by the sidewall LER. It should be noted that the off-state current variations in the gate LER is relatively smaller than the on-state current variations, because the fully depleted thin channel can effectively block the punch-through effect in spite of channel-length variations. For the same reason, the current variations induced by the fin LER with spacer patterning is the least.

The impact of channel height on the various performance metrics for different fin angles is summarized in Fig. 5.14. Because the effective channel width is increased as the channel height increases, the channel region can conduct more current at the same voltage. This means that there is a decrease in the threshold voltage. The off-state current is more sensitive to the channel height than the on-state current. In particular, the off-state current increases dramatically when the fin angle decreases because the gate-to-channel coupling becomes weak at the wide bottom of the channel region. A steeper increase in the off-state current leads to an increase in the SS. Additionally, the DIBL variations increase when the channel fin height increases. For all performance variations, except for the on-state current variations, the variation sensitivity has a minimum value when the fin angle is 90°. Therefore, a rectangular channel region is desirable in taller and narrower tri-gate MOSFETs.

While there is no line-width roughness (LWR) in the fin LER by spacer patterning, the tri-gate MOSFET still suffers strain variations (induced by the curvature of the channel region) from the LER in the range of 3 nm to 6 nm (in terms of the three-sigma value of LER amplitude). In [35], the longitudinal stress and the hole density of the p-type tri-gate MOSFET channel for misaligned and aligned sources/drains are presented. Similar to the erosional energy, which concentrates on the outside of a meandering stream, the stress concentrates on the outside of the curve. The average channel stress decreases by 12 and 8 % for misaligned and aligned sources/drains, respectively [35]. Thus, the carrier mobility is altered within the channel region because of the "fin LER"-induced strain variability. Furthermore, the band structure is varied depending on the strain, leading to a threshold voltage shift by tens of millivolts, which is a considerable effect in the sub-1 V operation. In addition, the fin LER has more effect on the channel length because most of the carriers flow along the fin curvature.

5.4 WFV in Tri-Gate MOSFET

It was verified that, regardless of the device size and the average grain size, the WFV in the planar bulk MOSFET can be easily and accurately estimated using the RGG plot [36]. Note that RGG refers to the ratio of average grain size to gate area. In the tri-gate MOSFET, the WFV-induced threshold voltage variation can be estimated using the RGG concept. However, the RGG should be modified in the

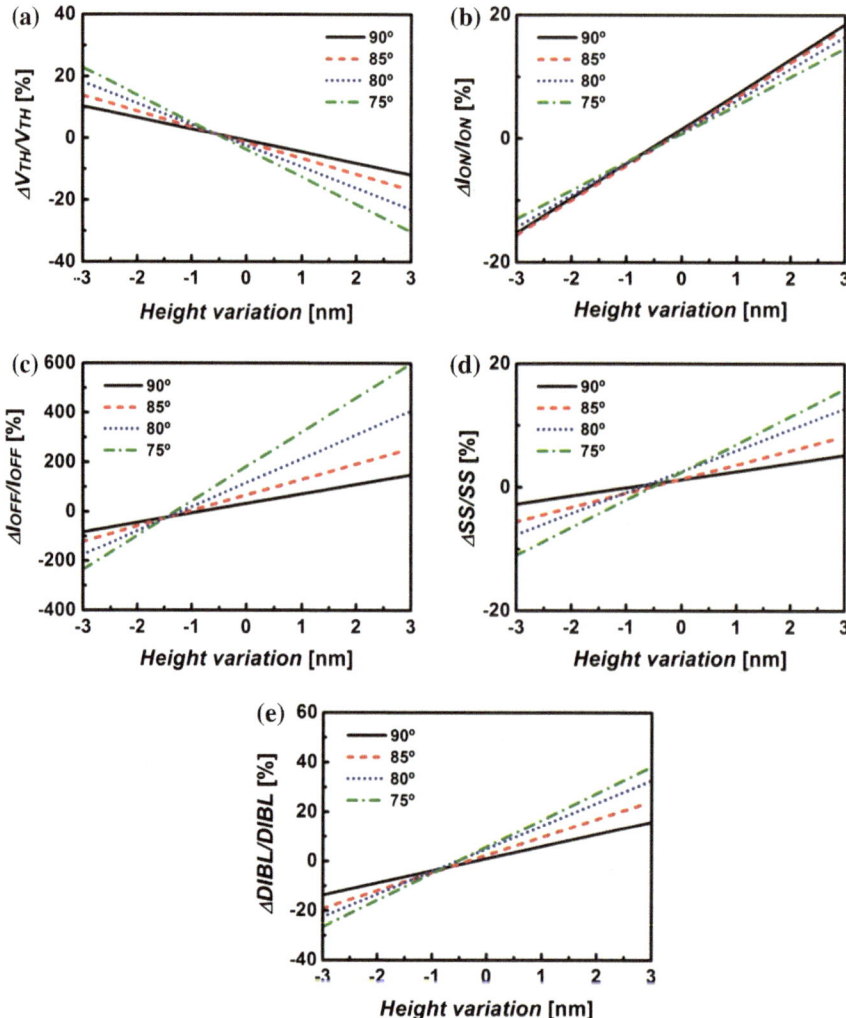

Fig. 5.14 Impact of channel height variation on **a** threshold voltage, **b** on-state current, **c** off-state current, **d** subthreshold slope, and **e** DIBL [16]

tri-gate MOSFET because the front and back gates simultaneously control the channel potential, as opposed to the single top gate in the planar bulk MOSFET. Hence, the definition of RGG in the tri-gate MOSFET should be modified as follows:

$$RGG = \frac{average\ grain\ size}{\sqrt{gate\ width + (2 \times gate\ height)}}$$

Figure 5.15 shows the schematic diagram, which describes the effect of the multiple gates. Because the front and back gates are associated with the formation of an inversion layer in the channel region, the two face-to-face gates mutually interact with each other while turning the transistor on or off. This effect, from the mutual interaction of the two face-to-face gates, can be taken into account during the RGG calculation by including imaginary gate areas. These gate areas derived from the interaction are composed of new grains with new probability and work function values, and these newly derived gate areas are called extended gate areas (EGAs) [17]. The addition of EGAs effectively widens the gate area, and therefore, the modified RGG for a tri-gate MOSFET with EGA should be defined as follows:

$$RGG|_{EGA} = \frac{average\ grain\ size}{\sqrt{gate\ width + (4 \times gate\ height)}}$$

The standard deviation of the work-function variation (WFV) in a tri-gate bulk MOSFET with TiN metal gate is plotted against RGG (see Fig. 5.16). A plot of the WFV, estimated using the original RGG concept, which is based on the conventional gate area of channel length × channel width, does not match well with the verified RGG plot. This is because the original RGG concept does not consider the

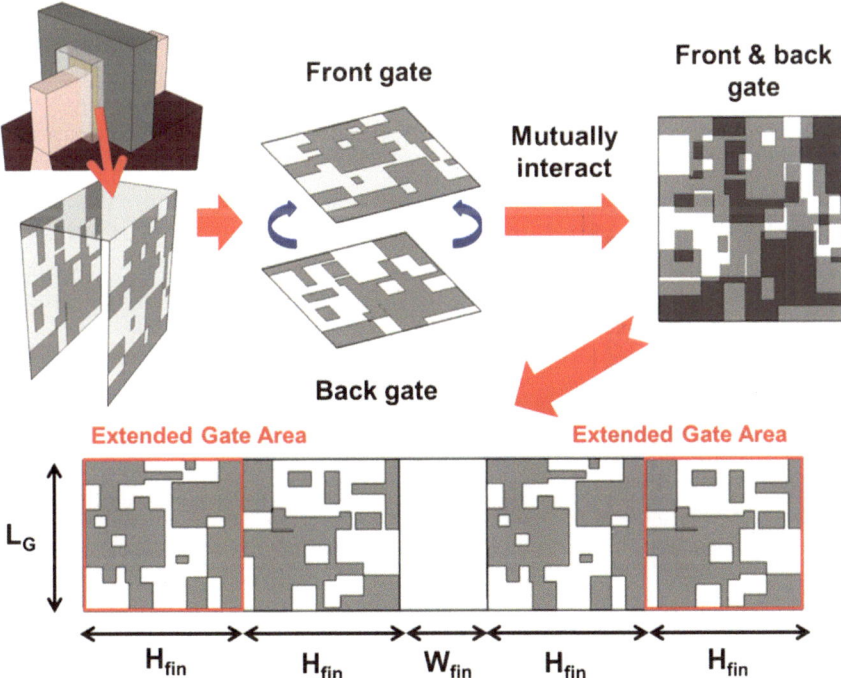

Fig. 5.15 Schematic diagram showing mutual interaction between the front gate and the back gate. The mutual interaction results in new grain generation [17]

Fig. 5.16 Standard deviation
of work function values
versus RGG [17]

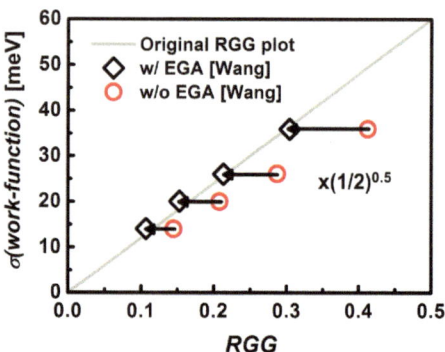

EGA effect. However, the quantity estimated using the modified RGG method (with the EGA) matches exactly with the previously verified RGG plot. Therefore, in order to preserve the original purpose/intention of using the RGG plot (i.e., to estimate the WFV for a given material by matching it to the original RGG plot, regardless of the device structure), the EGA should be considered for a tri-gate bulk MOSFET.

According to [36, 37], while the amount of WFV increases proportionally to RGG below a certain critical value, beyond the critical value, the WFV does not increases linearly and it begins to saturate. The critical value for planar bulk MOSFETs is 0.5 because the number of grains in the metal gate begins to saturate towards a value of 1 or 2 at an RGG of 0.5. An interesting fact is that, the critical point of the RGG plot for a tri-gate MOSFET with EGA is decreased by $(1/2)^{0.5}$ from 0.5 to 0.35 (see Fig. 5.17). This is because the total number of grains in the tri-gate MOSFET is modified by the EGA effect [17], even though the total number of grains in it is saturated in the range of 1–2 at an RGG of 0.5 (i.e., without EGA).

It should be noted that the purpose of considering the EGA effect in the RGG calculation is to compare the WFV between different device structures. If we do not consider the EGA effect while comparing the sensitivity of various device structures to WFV, we can reach the conclusion that the amount of WFV can be decreased by

Fig. 5.17 RGG plot for
planar bulk MOSFET and
tri-gate bulk MOSFET [17]

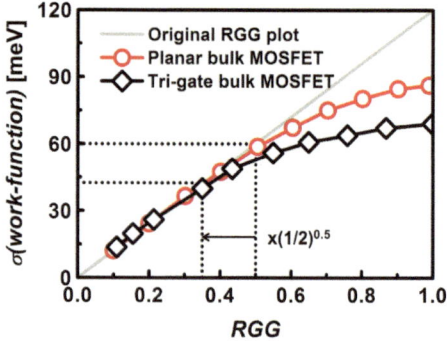

30 % in the tri-gate MOSFET, when compared to the planar bulk MOSFET [38]. In fact, the reduction of WFV in the tri-gate MOSFET originates from the 30 % increment in the gate area by the EGA effect. In short, the WFV suppression can be achieved in a simple manner by introducing a three-dimensional device structure without the use of a new metal gate material, and therefore, the double-gate device architecture represents a promising way for significantly suppressing the WFV in the future high-k metal-gate (HK/MG) CMOS technology.

However, the variation of the electrical characteristics of the tri-gate MOSFET based on the shape of the channel region (e.g., tapered vs. rectangular channel region) [39, 40], results in the alteration of the impact of WFV on the threshold voltage. More specifically, the shape of the current flow in the channel region determines whether the EGA effect should be included in the RGG computation, or not [18]. The fin-shaped channel region in the tri-gate MOSFET was developed, and it evolved from the tapered shape in 2011 to the rectangular shape in 2014 [13, 33]. Figure 5.18 shows the 3-D bird's-eye view and the cross-sectional view of the rectangular and tapered tri-gate MOSFETs. Their input characteristics are shown in Fig. 5.19. For identical bottom widths, a narrower top width in the tapered channel region results in a better gate-to-channel controllability [39] and an improved SS and DIBL [39, 40]. In addition, the different channel shapes induce different current-flow shapes in the channel region: surface current flow in the rectangular channel region versus

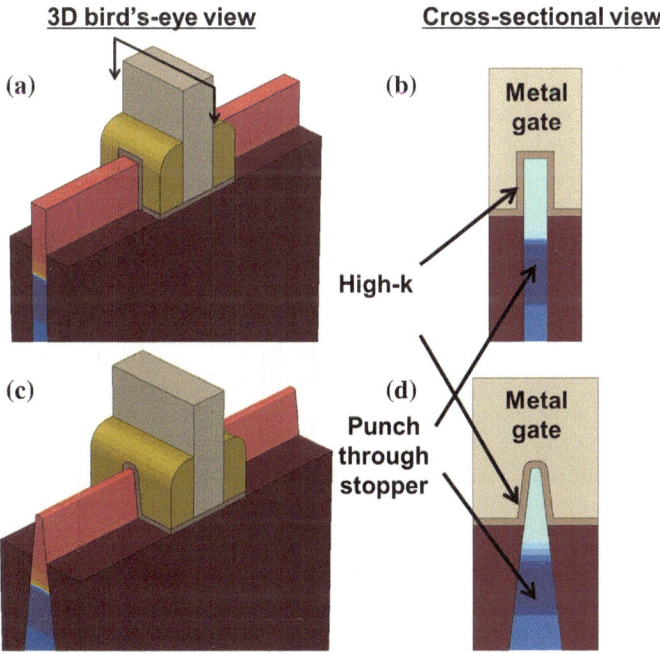

Fig. 5.18   Three-dimensional (3-D) bird's-eye views of **a** rectangular and **c** tapered tri-gate MOSFET. **b** Cross-sectional view of **a**. **d** Cross-sectional view of **c**

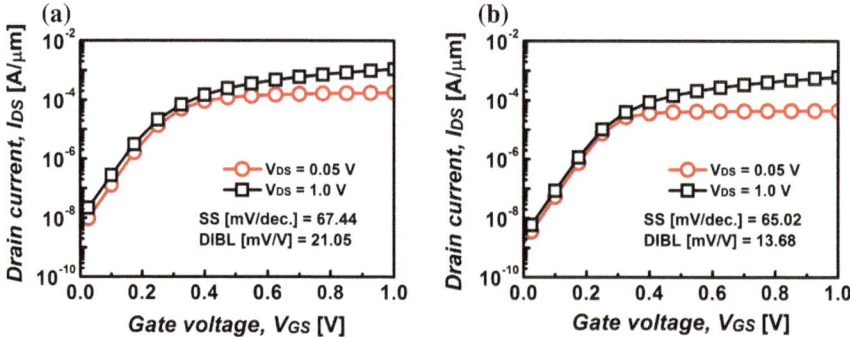

Fig. 5.19 Input transfer characteristics of **a** rectangular and **b** tapered tri-gate MOSFET [18]

bulky current flow in the tapered channel region (see Fig. 5.20). The different current-flow shapes are determined at threshold voltage [18]. When considering the EGA effect, the amount of the threshold voltage variations in the rectangular tri-gate MOSFET is matched to the previous RGG plot for the tri-gate MOSFET (see Fig. 5.21a). However, when excluding the EGA effect, the amount of WFV-induced threshold voltage variations in tapered tri-gate MOSFET is matched to the previous RGG plot for the tri-gate MOSFET (see Fig. 5.21b). The simulation results, with the EGA, largely deviate from those of the previous RGG plots for the tri-gate MOSFET. Thus, one can conclude that EGA effects should be included in the RGG calculation only when the channel region has two isolated surface current-flow shapes (see Fig. 5.20a). In order to verify this conclusion, the WFV-induced threshold voltage variations in a rectangular tri-gate MOSFET with a single bulky current-flow shape (see Fig. 5.22a) and in a tapered tri-gate MOSFET with two independent surface current-flow shapes (see Fig. 5.22b) were estimated. As shown in Fig. 5.23, the simulation results for the rectangular (tapered) tri-gate MOSFET without (with) EGA

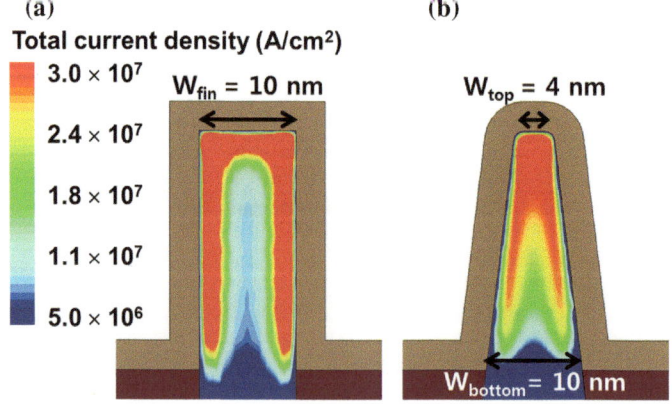

Fig. 5.20 Total current density of **a** rectangular and **b** tapered tri-gate MOSFET

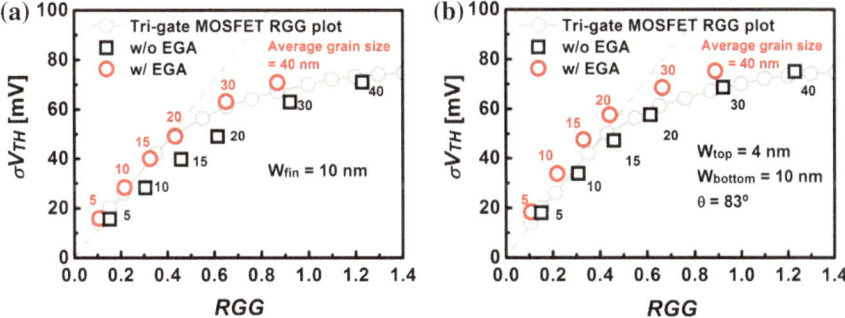

Fig. 5.21 Standard deviation of WFV-induced threshold voltage variations versus RGG in **a** rectangular and **b** tapered tri-gate MOSFET [18]

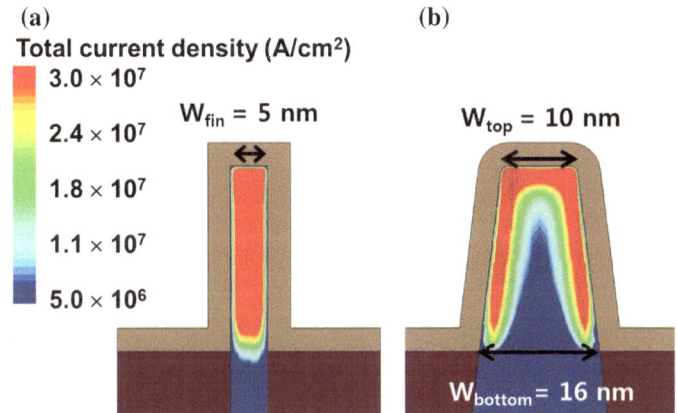

Fig. 5.22 Total current density of **a** rectangular and **b** tapered tri-gate MOSFET

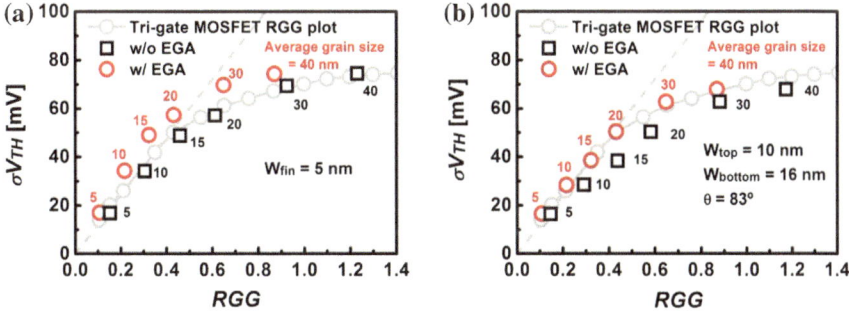

Fig. 5.23 Standard deviation of WFV-induced threshold voltage variation versus RGG in **a** rectangular and **b** tapered tri-gate MOSFET [18]

match exactly with the previous RGG plots for the tri-gate MOSFET [17]. Finally, although the tapered tri-gate MOSFET is slightly more favorable in terms of improving the SS and DIBL, we should design the tri-gate MOSFET to have (i) rectangular channel region and (ii) surface current-flow shape, in order to decrease the impact of WFV-induced threshold voltage variations by 30 % (i.e., by exploiting the EGA effect).

References

1. Packan P, Akbar S, Armstrong M, Bergstrom D, Brazier M, Deshpande H, Dev K, Ding G, Ghani T, Golonzka O, Han W, He J, Heussner R, James R, Jopling J, Kenyon C, Lee S-H, Liu M, Lodha S, Mattis B, Murthy A, Neiberg L, Neirynck J, Pae S, Parker C, Pipes L, Sebastian J, Seiple J, Sell B, Sharma A, Sivakumar S, Song B, St. Amour A, Tone K, Troeger T, Weber C, Zhang K, Luo Y, Natarajan S (2009) High performance 32 nm logic technology featuring 2nd generation high-k+ metal gate transistors. In: Proceedings of IEEE IEDM, pp 1–4

2. Asenov A, Kaya S, Brown AR (2003) Intrinsic parameter fluctuations in decananometer MOSFETs introduced by gate line edge roughness. IEEE Trans Electron Devices 50(5):1254–1260

3. Asenov A (1998) Random dopant induced threshold voltage lowering and fluctuations in sub-0.1 μm MOSFETs: a 3-D "atomistic" simulation study. IEEE Trans Electron Devices 45(12):2505–2513

4. Brown AR, Roy G, Asenov A (2007) Poly-Si-Gate-related variability in decananometer MOSFETs with conventional architecture. IEEE Trans Electron Devices 54(11):3056–3063

5. Yau LD (1974) A simple theory to predict the threshold voltage for short-channel IGFET's. Solid-State Electronics 17(10):1059–1063

6. Yan R-H, Ourmazd A, Lee KF (1992) Scaling the Si MOSFET: from bulk to SOI to bulk. IEEE Trans Electron Devices 39(7):1704–1710

7. Hanafi HI, Noble WP, Bass RS, Varahramyan K, Lii Y, Dally AJ (1993) A model for anomalous short-channel behavior in submicron MOSFET's. IEEE Electron Device Lett 14(12):575–577

8. Sadana D, Acovic A, Shahidi G, Hanafi H, Warren A, Grutzrnacher D, Cardone F, Sun J, Davari B (1992) Enhanced short-channel effects in NMOSFET's due to boron redistribution induced by arsenic source and drain implant. In: IEDM technical digest, pp 849–852

9. Gwoziecki R, Skotnicki T, Bouillon P, Gentil P (1999) Optimization of V_{TH} roll-off in MOSFET's with advanced channel architecture—retrograde doping and pockets. IEEE Trans Electron Devices 46(7):1551–1561

10. Troutman RR (1979) VLSI limitations from drain-induced barrier lowering. IEEE Trans Electron Devices 26(4):461–469

11. Fjeldly TA, Shur M (1993) Threshold voltage modeling and the subthreshold regime of operation of short-channel MOSFET's. IEEE Trans Electron Devices 40(1):137–145

12. Chamberlain SG, Ramanan S (1986) Drain-induced barrier-lowering analysis in VSLI MOSFET devices using two-dimensional numerical simulations. IEEE Trans Electron Devices 33(11):1745–1753

13. Auth C, Allen C, Blattner A, Bergstrom D, Brazier M, Bost M, Buehler M, Chikarmane V, Ghani T, Glassman T, Grover R, Han W, Hanken D, Hattendorf M, Hentges P, Heussner R, Hicks J, Ingerly D, Jain P, Jaloviar S, James R, Jones D, Jopling J, Joshi S, Kenyon C, Liu H, McFadden R, McIntyre B, Neirynck J, Parker C, Pipes L, Post I, Pradhan S, Prince M, Ramey S, Reynolds T, Roesler J, Sandford J, Seiple J, Smith P, Thomas C, Towner D,

Troeger T, Weber C, Yashar P, Zawadzki K, Mistry K (2012) A 22 nm high performance and low-power CMOS technology featuring fully-depleted tri-gate transistors, self-aligned contacts and high density MIM capacitors. In: Symposium on VLSI technical digest, pp 131–132

14. Takeuchi K, Fukai T, Tsunomura T, Putra AT, Nishida A, Kamohara S, Hiramoto T (2007) Understanding random threshold voltage fluctuation by comparing multiple fabs and technologies. In: IEDM technical digest, pp 467–470

15. Wang X, Roy G, Saxod O, Bajolet A, Juge A, Asenov A (2012) Simulation study of dominant statistical variability sources in 32-nm high-k/metal gate CMOS. IEEE Electron Device Lett 33(5):643–645

16. Huang W-T, Li Y (2014) The impact of fin/sidewall/gate line edge roughness on trapezoidal bulk FinFET devices. In: IEEE SISPAD, pp 281–284

17. Nam H, Shin C (2013) Study of high-k/metal-gate work-function variation in FinFET: the modified RGG concept. IEEE Electron Devices Lett. 34(12):1560–1562

18. Nam H, Shin C (2014) Impact of current flow shape in tapered (versus rectangular) FinFET on threshold voltage variation induced by work-function variation. IEEE Trans Electron Devices 61(6):2007–2011

19. Shin C, Sun X, Liu T-JK (2009) Study of random-dopant-fluctuation (RDF) effects for the trigate bulk MOSFET. IEEE Trans Electron Devices 56(7):1538–1542

20. Varadarajan V, Smith L, Balasubramanian S, King Liu T-J (2006) Multigate FET design for tolerance to statistical dopant fluctuations. In: Proceedings of silicon nanoelectronics workshop, pp 137–138

21. Mizuno T, Okamura J-I, Toriumi A (1994) Experimental study of threshold voltage fluctuation due to statistical variation of channel dopant number in MOSFETs. IEEE Trans Electron Devices 41(11):2216–2221

22. Sun X, Liu Q, Moroz V, Takeuchi H, Gebara G, Wetzel J, Ikeda S, Shin C, King Liu T-J (2008) Tri-gate bulk MOSFET design for CMOS scaling to the end of the roadmap. IEEE Electron Device Lett 29(5):491–493

23. Su H-W, Li Y, Chen Y-Y, Chen C-Y, Chang H-T (2012) Drain-induced-barrier lowering and subthreshold swing fluctuations in 16-nm-gate bulk FinFET devices induced by random discrete dopants. In: IEEE device research conference (DRC), pp 109–110

24. Miyamura M, Nagumo T, Takeuchi K, Takeda K, Hane M (2008) Effects of drain bias on threshold voltage fluctuation and its impact on circuit characteristics. In: Proceedings of IEEE IEDM, pp 1–4

25. Varadarajan V, Smith L, Liu T-JK (2009) FinFET design for tolerance to statistical dopant fluctuations. IEEE Trans Nanotechnol 8(3):375–378

26. Wang X, Brown AR, Cheng B, Asenov A (2011) Statistical variability and reliability in nanoscale FinFETs. In: Proceedings of IEEE IEDM, pp 5.4.1–5.4.4

27. Baravelli E, Dixit A, Rooyackers R, Jurczak M, Speciale N, Meyer KD (2007) Impact of line-edge roughness on FinFET matching performance. IEEE Trans Electron Devices 54 (9):2466–2474

28. Cheng Q, You J, Chen Y (2014) Correlating FinFET device variability to the spatial fluctuation of fin width. Microelectron Eng 119:53–60

29. Agrawal N, Kimura Y, Arghavani R, Datta S (2013) Impact of transistor architecture (bulk planar, trigate on bulk, ultrathin-body planar SOI) and material (silicon or III–V Semiconductor) on variation for logic and SRAM applications. IEEE Trans. Electron Device 60(10):3298–3304

30. Leung G, Chui CO (2013) Interactions between line edge roughness and random dopant fluctuation in nonplanar field-effect transistor variability. IEEE Trans Electron Device 60 (10):3277–3284

31. Chen C-H, Li Y, Chen C-Y, Chen Y-Y, Hsu S-C, Huang W-T, Chu S-Y (2011) Mobility model extraction for surface roughness of SiGe along (110) and (100) Orientations in HKMG bulk FinFET devices. Microelectron Eng 109:357–359

32. Lin C-H, Haensch W, Oldiges P, Wang H, Williams R, Chang J, Guillorn M, Bryant A, Yamashita T, Standaert T, Bu H, Leobandung E, Khare M (2011) Modeling of width-quantization-induced variations in logic FinFETs for 22 nm and beyond. In: VLSI symposium on technical digest, pp 16–17

33. Natarajan S, Agostinelli M, Akbar S, Bost M, Bowonder A, Chikarmane V, Chouksey S, Dasgupta A, Fischer K, Fu Q, Ghani T, Giles M, Govindaraju S, Grover R, Han W, Hanken D, Haralson E, Haran M, Heckscher M, Heussner R, Jain P, James R, Jhaveri R, Jin I, Kam H, Karl E, Kenyon C, Liu M, Luo Y, Mehandru R, Morarka S, Neiberg L, Packan P, Paliwal A, Parker C, Patel P, Patel R, Pelto C, Pipes L, Plekhanov P, Prince M, Rajamani S, Sandford J, Sell B, Sivakumar S, Smith P, Song B, Tone K, Troeger T, Wiedemer J, Yang M, Zhang K (2014) A 14 nm logic technology featuring 2nd-generation FinFET, air-gapped interconnects, self-aligned double patterning and a 0.0588 μm^2 SRAM cell size. In: Proceedings of IEEE IEDM, pp 3.7.1–3.7.3

34. Degroote B, Rooyackers R, Vandeweyer T, Collaert N, Boullart W, Kunnen E, Shamiryan D, Wouters J, Van Puymbroeck J, Dixit A, Jurczak M (2007) Spacer defined FinFET: active area patterning of sub-20 nm fins with high density. Microelectron Eng 84:609–618

35. Choi M, Moroz V, Smith L, Penzin O (2012) 14 nm FinFET stress engineering with epitaxial SiGe source/drain. In: Silicon-germanium technology and device meeting (ISTDM), pp 1–2

36. Nam H, Shin C (2013) Study of high-k/metal-gate work-function variation using Rayleigh distribution. IEEE Electron Device Lett 34(4):532–534

37. Nam H, Shin C (2013) Comparative study in work-function variation: Gaussian vs. Rayleigh distribution for grain size. IEICE Electron Expr 10(9):20130109

38. Matsukawa T, O'uchi S, Ishikawa Y et al (2009) Comprehensive analysis of variability sources of FinFET characteristics. In: Proceedings of symposium on VLSI technology, pp 118–119

39. Moroz V (2012) FinFET structure design and variability analysis enabled by TCAD. Available http://www.embedded.com/print/4398011

40. Ko MD, Sohn CW, Baek CK, Jeong YH (2013) Study on a scaling length model for tapered tri-gate FinFET based on 3-D simulation and analytical analysis. IEEE Trans Electron Devices 60(9):2721–2727

Chapter 6
Quasi-Planar Trigate (QPT) Bulk MOSFET

6.1 QPT Bulk MOSFET

Following the Moore's Law, the continuous miniaturization of complementary metal oxide semiconductor (CMOS) transistors for improving the performance of integrated circuit (IC) chips has intensified the process-induced random variation [i.e., the threshold voltage (V_T) variation caused by line-edge roughness (LER), random dopant fluctuation (RDF), and work-function variation (WFV)]. The V_T variation became significant in the 45-nm CMOS technology node and beyond. Specifically, the process-induced random variation in V_T provoked the non-negligible V_T mismatch in the bit cell of the static random access memory (SRAM) because the bit cell employed state-of-the-art transistors (i.e., the smallest transistors in physical size) [1]. The LER-/RDF-/WFV-induced V_T mismatch in SRAM cells obscured the scaling of the operating voltage, resulting in increased power density in the SRAM [2]. Hence, a variation-immune device design to minimize the process-induced V_T variation as well as to improve the gate-to-channel control for reducing the short channel effects (SCEs) becomes important, whereupon some alternative designs of thin-body device structure are considered (e.g., fully depleted silicon-on-insulator (FD-SOI) MOSFET [3], FinFET, and Tri-gate MOSFET [4]). However, these advanced device structures impose certain additional expenses to use the silicon-on-insulator (SOI) substrate and have relatively complicated fabrication processes as compared to the conventional planar bulk MOSFETs. To address the aforementioned issues, a quasi-planar trigate (QPT) bulk CMOS transistor was designed [5, 6] to alleviate the process-induced random variation with reasonable cost and simple fabrication steps. For example, the QPT bulk MOSFET includes the conventional retrograde channel doping profile unlike the thin-body structures (i.e., FD-SOI MOSFET, FinFET, and Tri-gate MOSFET) in order to restrain leakage current as well as to improve the gate-to-channel controllability.

© Springer Science+Business Media Dordrecht 2016   91
C. Shin, *Variation-Aware Advanced CMOS Devices and SRAM*,
Springer Series in Advanced Microelectronics 56,
DOI 10.1007/978-94-017-7597-7_6

6.2   Fabrication of a QPT Bulk MOSFET

A 28-nm CMOS logic technology was used to experimentally verify the improved performance characteristics of the QPT bulk MOSFET in an SRAM cell. Figure 6.1 shows the fabrication sequence of the front-end-of-line QPT bulk MOSFET. In order to fabricate the QPT bulk MOSFET, (100) epi-silicon wafers were prepared with $\langle 110 \rangle$ channel orientation. After the shallow trench isolation (STI) formation process, the steps for n/p well formation and ion implantation for threshold voltage (V_T) adjustment were followed. Further, rapid thermal annealing (RTA) was performed to remove the damages (defects) caused by the ion implantation process. Afterwards, residual sacrificial oxide was etched out using dilute hydrofluoric acid (DHF) and the upper part of the STI oxide was removed in order to create the gate stack of the quasi-planar structure. The baseline MOSFET (i.e., the planar bulk MOSFET) was also fabricated using a shorter DHF dip in order not to form the quasi-planar device structure. Then, a gate oxide layer of 1.45 nm was formed using the plasma nitridation method to achieve a high quality gate oxide layer and an intrinsic polycrystalline silicon layer of 70 nm was deposited in succession. A double-patterning/double-etching (2P2E) technique using 193-nm immersion lithography was used to define the gate electrodes and the 0.149 μm^2 six-transistor (6-T) SRAM bit cells. After the 2P2E patterning technique, the processes for pocket ion implantation and gate-sidewall spacers were carried out. Subsequently, the ion implantation for source/drain regions was performed and then a rapid thermal process (RTP), followed by laser spike annealing (LSA) was used to enhance the electrical conductivity in the source/drain regions. Further, a nickel silicidation (NiSi) layer was deposited. Thereafter, stress engineering using dual contact etch stop layer (CESL) was applied for improving the mobility of both the n-type and p-type transistors. In this process, compressive stress and tensile stress were applied to the PMOS and NMOS, respectively, using a SiN_x layer, which was deposited by the plasma-enhanced chemical vapor deposition (PECVD) method. In succession, in the back-end-of-line process, the interlayer dielectric (ILD) oxide layer was

Fig. 6.1 Process flow of
quasi-planar transistor
(QPT) for an SRAM array

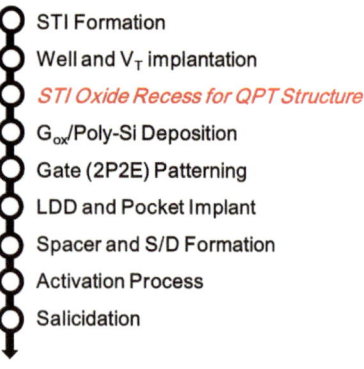

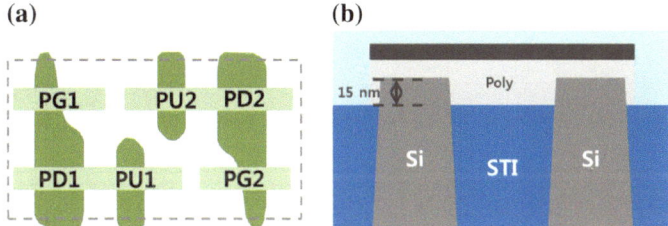

Fig. 6.2 a Illustrated top view of the fabricated QPT-based 0.149-μm² SRAM cell and **b** illustrated cross-sectional view of the QPT SRAM cell along with the poly-silicon gate. Note that the *upper part* of the STI oxide was etched out by 15 nm

deposited on top of the silicon nitride capping layer. Subsequently, the contact hole definition, tungsten plug formation, and chemical mechanical planarization (CMP) process were performed. Finally, a copper metal layer was deposited for transistor interconnection.

In order to fabricate the QPT bulk transistors and the QPT-based 6-T SRAM arrays, a standard test-chip mask set was used. This mask includes 2500 cells per device-under-test (DUT). Figure 6.2a illustrates the plan view of the QPT-based 6-T SRAM cell and Fig. 6.2b shows the illustrated image of an SRAM cell along with the poly-silicon gate electrode.

6.3 Results and Discussion

6.3.1 Improved Performance in QPT Bulk MOSFET (Vs. Conventional MOSFET)

Because of the quasi-planar structure created by the removal process of the STI oxide region, the effective channel width of the QPT bulk MOSFET is increased and the gate controllability is improved. As shown in Fig. 6.3, an improved on-state current (I_{ON} or *Idsat*) of the QPT bulk MOSFET is achieved (but with the off-state current (I_{OFF}) of the conventional planar bulk MOSFET). In addition, the low pocket doping process leads to lower threshold voltage with higher effective mobility in the channel region, which also induces higher I_{ON} (or *Idsat*). Owing to the additional channel width contributed by the sidewall gate regions, the n-type pass-gate (PG) transistors shows 2.4 times the on-state current than that of the baseline MOSFET. Furthermore, there is a greater performance improvement in the n-type pull-down (PD) transistors (i.e., 2.1 times the I_{ON}). Besides the performance improvement provided by the n-type transistors, the p-type transistors (i.e., the pull-up (PU) transistors in the SRAM cell) evidently achieve the greatest

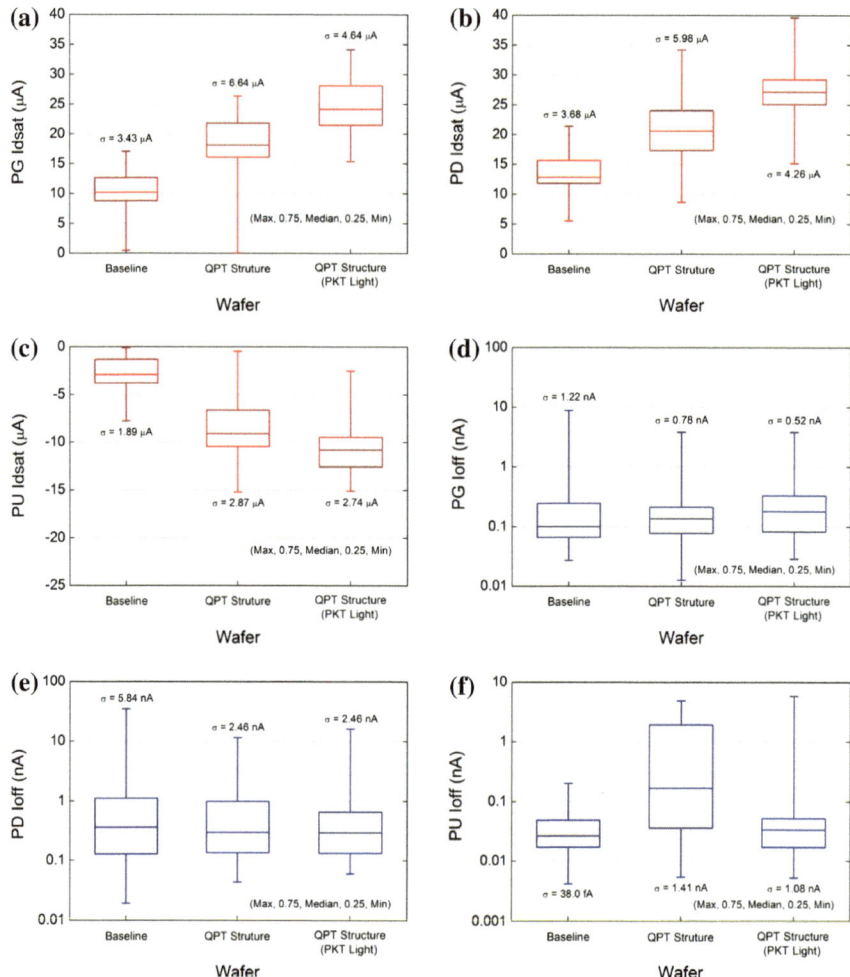

Fig. 6.3 Comparison of a baseline MOSFET (Baseline), a quasi-planar transistor with a 15 nm recess (QPT Structure), and a quasi-planar transistor with 15 nm recessed oxide and lighter pocket doping concentration (QPT Structure–PKT Light) in SRAM cells. **a** *Idsat* of NMOS pass-gate (PG), **b** *Idsat* of NMOS pull-down (PD), **c** *Idsat* of PMOS pull-up (PU), **d** *Ioff* of NMOS PG, **e** *Ioff* of NMOS PD, and **f** *Ioff* of PMOS PU. This data has been adapted from [11]

enhancement in device performance. The PMOS demonstrated a performance enhancement of 4.5×, primarily caused by (i) the narrow layout width of the PU device (i.e., increased stress effects) and (ii) the improved hole mobility for (110) sidewall surfaces of the channel region (note that the electron mobility is degraded for the same surface directions [7]).

6.3.2   Suppressed V_T Variation by the QPT Structure

Figure 6.4 shows the V_T variation characteristics of the PG/PD/PU transistors in an SRAM cell. For all the three transistors in the SRAM bit cell, it is observed that the subthreshold slope becomes steeper and thereby V_T is lowered because of the increased gate controllability in the QPT bulk MOSFETs. In the V_T statistics, a little larger V_T variation is observed for the QPT-based PG and PD transistors. This is because the standard dose for pocket implantation in the early 28 nm technology for NMOS transistors is slightly higher than that of the PMOS transistors (i.e., the impact of the RDF on NMOS gate sidewalls is relatively higher than that on PMOS's gate sidewalls). Thus, in order to remove the aforementioned undesirable effect, a lighter pocket implant dose for the QPT-based PG and PD transistors is used. The lower dose alleviates the V_T variation but with a comparable off-state leakage current, as shown in Fig. 6.4. On the other hand, because the standard pocket implant dose for PMOS is relatively low, the RDF issue in the PMOS is not

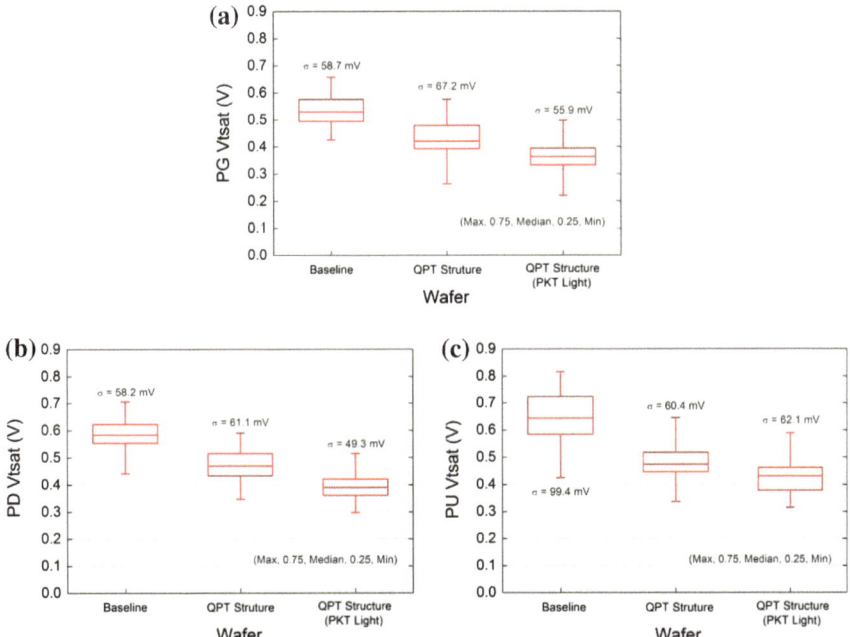

Fig. 6.4 Comparison of the saturation threshold voltage (V_{tsat}) of a baseline MOSFET (Baseline), a quasi-planar transistor with a 15-nm recess (QPT Structure), and a quasi-planar transistor with a 15-nm recess and light pocket doping concentration (QPT Structure–PKT Light) in SRAM cells. **a** V_{tsat} of NMOS pass-gate (PG), **b** V_{tsat} of NMOS pull-down (PD), and **c** V_{tsat} of PMOS pull-up (PU). This data has been adapted from [11]

(a)   **(b)**

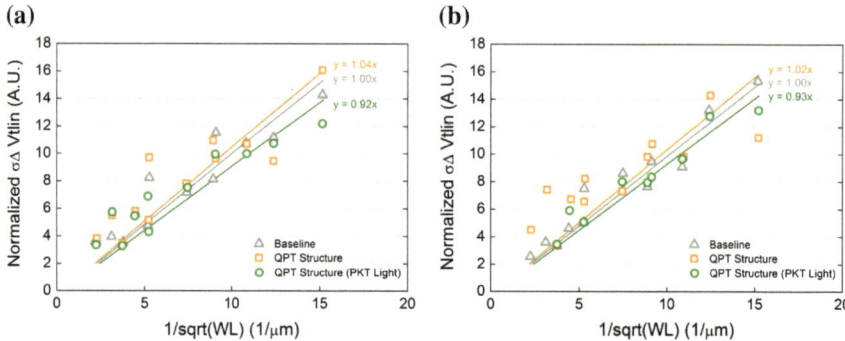

Fig. 6.5 Comparison of Pelgrom's plots for **a** N-type MOS devices and **b** P-type MOS devices. Note that the drawn channel width is in the range of 120 nm to 1 μm and the drawn channel length is in the range of 36 nm–0.2 μm. This data has been adapted from [11]

considered as a significant gating issue. Thus, the V_T variation can be improved with the removal process of the STI oxide, owing to the advanced electrostatic coupling effect of the quasi-planar device structure. However, it is noteworthy that the V_T variation for the PMOS is slightly worsened because of the increased short channel effect with the reduced pocket implant dose. Therefore, the optimization of the channel and pocket doping concentration is needed for improving the magnitude of V_T variation as well as for satisfying the requirement of corners (i.e., slow-slow (SS), slow-fast (SF), fast-slow (FS), and fast-fast (FF) corners).

Figure 6.5 shows the Pelgrom's plot [8] for n-/p-type QPT bulk MOSFETs, which explicitly illustrates the trend of increase in the V_T variation with the decreasing channel area. In the Pelgrom's plot, the Pelgrom's coefficient (i.e., A_{VT}) is lowered for the quasi-planar transistors with a lower pocket implant dose (i.e., by 8 and 7 % for NMOS and PMOS, respectively).

6.3.3 Improved Short Channel Effect in the QPT Bulk MOSFET

Figure 6.6 shows the measured saturation threshold voltage (V_{tsat}) for various channel lengths of a quasi-planar MOSFET, which has a drawn channel width of 0.25 μm. In this plot, an improved V_T roll-off is observed for the quasi-planar structure because of better channel controllability. Especially, the V_T is considerably well-maintained for the NMOS devices. Further, even if the lighter pocket doping process was adopted, a reasonable short channel effect would be retained.

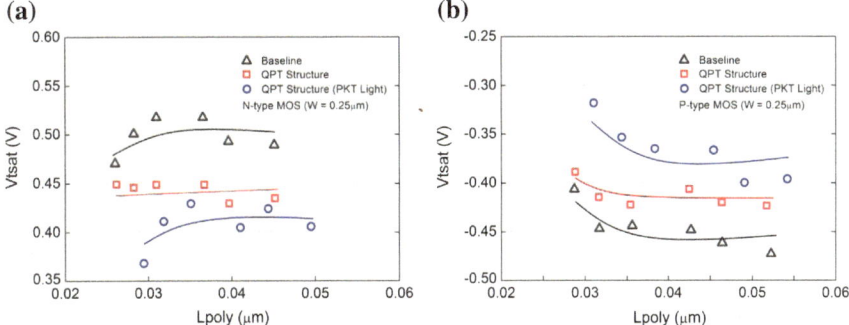

Fig. 6.6 V_{tsat} values according to the gate length for **a** NMOS and **b** PMOS devices with a gate width of 0.25 μm. This data has been adapted from [11]

6.3.4   Increased Narrow Width Effect in the QPT Bulk MOSFET

The reverse narrow width effect (RNWE) indicates that V_T becomes lower as the channel width gets narrower. The increased RNWE in the QPT structure originates from the fringing electric fields between the gate electrode margin and the channel's sidewalls. Therefore, as shown in Fig. 6.7, V_T roll-off in the quasi-planar structure becomes severe, but the average V_T is lowered with the recessed STI oxide region because of the intensified channel controllability.

As discussed in [9], wider channel width of transistor should be divided by less than or equal to $2 \times L_G$, where L_G is the channel length, in order to maximize the layout efficiency of the quasi-planar device structure. In order to form a highly uniform channel width, a double-patterning process can be adopted [10]. This

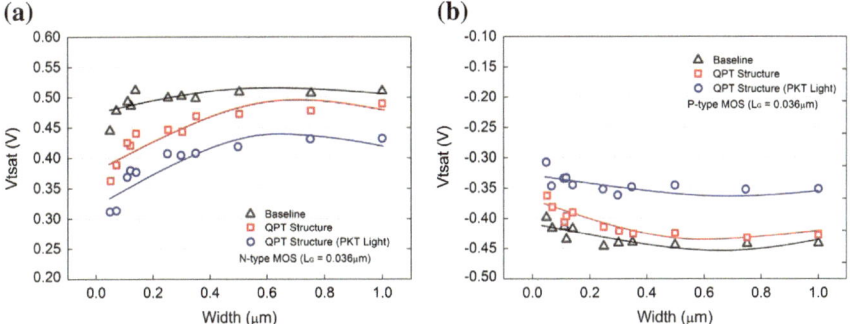

Fig. 6.7 V_{tsat} versus the channel width for a given channel length of 36 nm for **a** NMOS devices and **b** PMOS devices. Note that the average V_{tsat} value is lowered with the QPT device structure. This data has been adapted from [11]

process can achieve the uniformly segmented channel regions without employing a high-aspect-ratio isolation process.

6.3.5 A Compact Model for the QPT Bulk MOSFET

As the state-of-the-art transistors including FinFET and FD-SOI MOSFET, it is confirmed that the compact model of the QPT bulk MOSFET is compatible to the existing compact models of MOSFET for IC circuits. Accordingly, the BSIM4.6 compact model developed by the University of California Berkeley (Device group) was used. It was calibrated based on the electrical properties of the QPT bulk MOSFET. As shown in Fig. 6.8, the compact model is appropriately matched to the characteristics of the quasi-planar MOSFETs that include body effect. This model shows another strong point of the QPT bulk technology, i.e., it can demonstrate the adaptive body biasing for dynamic optimization between performance and energy efficiency.

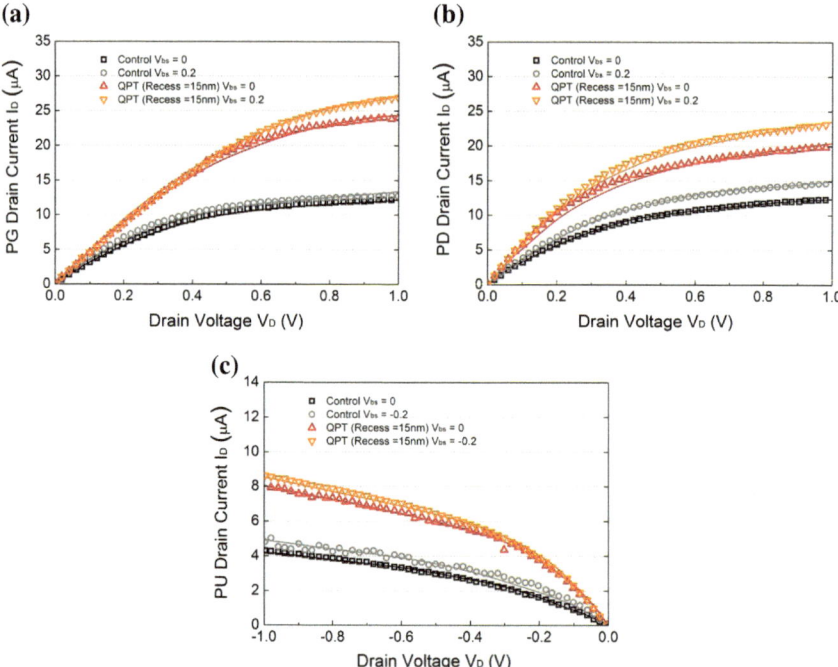

Fig. 6.8 Comparison of the output characteristic curve (i.e., *drain current* vs. *drain voltage* for a given gate voltage). **a** NMOS pass-gate (PG), **b** NMOS pull-down (PD), and **c** PMOS pull-up (PU). The *symbols* correspond to the measured data and the *solid lines* indicate the compact model data. Note that the applied gate voltage (V_{GS}) is 1.0 V. This data has been adapted from [11]

6.4   Benefits of the Quasi-Planar Bulk CMOS Technology for 6T-SRAM

6.4.1   Yield Enhancement in the QPT-Based 6-T SRAM Bit Cells

With the use of QPT bulk MOSFETs employing the early 28-nm CMOS technology, the QPT-based SRAM yield evaluated by the 3-sigma/median value of the static noise margin (SNM) and write noise margin (WNM) is slightly decreased because of the increased V_T variation for NMOS (see Fig. 6.9a, b). However, as shown in Fig. 6.9a, the improved variation characteristics for quasi-planar technology are confirmed when a lighter pocket doping is used. Although the drive current for PU transistors is improved with the quasi-planar structure, the nominal SNM becomes worse because of the degraded beta ratio in an SRAM bit cell. Thus, the 3-sigma/median value is deteriorated, but a relatively smaller value of the 3-sigma/median is obtained because of the improved drive current of the PU transistor. In addition, as shown in Fig. 6.9b, the 3-sigma/median of the nominal WNM is slightly increased in spite of the abrupt increment of sigma value originating from the increased V_T variation of the PG.

6.4.2   Scaling of the Power Supply Voltage (V_{DD})

In order to reduce the ever-increasing power density in an IC chip, the reduction of power supply voltage (V_{DD}) is significantly considered. However, the decrease in the overdrive voltage (i.e., $V_{DD} - V_T$) can lead to an increase in the variability, resulting in a worse SRAM yield.

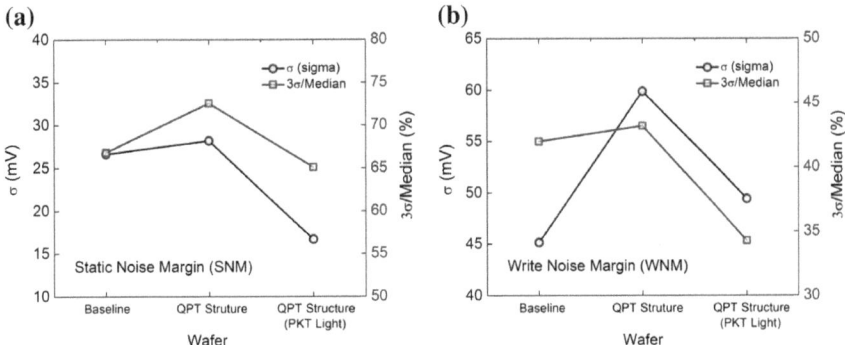

Fig. 6.9 σ and 3σ/median values for **a** SNM and **b** WNM with $V_{DD} = 1.0$ V. This data has been adapted from [11]

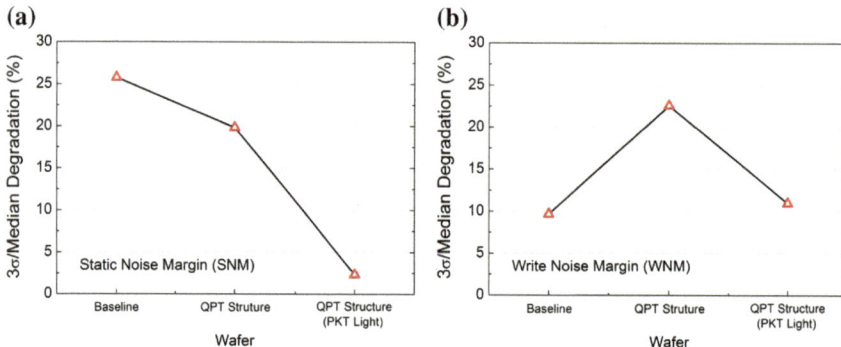

Fig. 6.10 3σ/median degradation with the reduced power supply voltage (V_{DD}) for **a** SNM and **b** WNM. This data has been adapted from [11]

Figure 6.10a shows the degradation of SNM when V_{DD} is reduced from 1.0 to 0.8 V. The 3-sigma/median degradation is improved with the quasi-planar structure. Especially, a considerable improvement in the 3-sigma/median is observed for a light pocket doping concentration, without a significant degradation in the WNM yield (see Fig. 6.10b). If the optimization process of pocket implantation dose is conducted separately for the NMOS/PMOS transistors, the yield of SNM and WNM can be effectively controlled with V_{DD} scaling.

References

1. Bowman K et al (2000) Impact of extrinsic and intrinsic parameter fluctuations on CMOS circuit performance. Solid-State Circ IEEE J 35(8):1186–1193
2. Nii K et al (2008) A 45-nm single-port and dual-port SRAM family with robust read/write stabilizing circuitry under DVFS environment. In: Proceedings of IEEE symposium on VLSI circuits IEEE, NY
3. Fenouillet-Beranger C et al (2009) FDSOI devices with thin BOX and ground plane integration for 32 nm node and below. Solid-State Electron 53(7):730–734
4. Kawasaki H et al (2008) Demonstration of highly scaled FinFET SRAM cells with high-κ/metal gate and investigation of characteristic variability for the 32 nm node and beyond. In: 2008 IEEE international electron devices meeting, IEDM 2008. IEEE, NY
5. Tsai CH et al (2010) Segmented tri-gate bulk CMOS technology for device variability improvement. In: Proceedings of the IEEE international symposium on VLSI technology systems and applications (VLSI-TSA)
6. Shin C et al (2010) Tri-gate bulk CMOS technology for improved SRAM scalability. In: Proceedings of the European IEEE conference on solid-state device research (ESSDERC)
7. Chang L, Ieong M, Yang Min (2004) CMOS circuit performance enhancement by surface orientation optimization. Electron Devices IEEE Trans 51(10):1621–1627
8. Pelgrom MJM, Duinmaijer ACJ, Welbers APG (1989) Matching properties of MOS transistors. IEEE J Solid-State Circ 24(5):1433–1439

9. Sun X et al (2008) Tri-gate bulk MOSFET design for CMOS scaling to the end of the roadmap. Electron Device Lett IEEE 29(5):491–493

10. US Patent 7190.050

11. Shin C et al (2011) Quasi-planar bulk CMOS technology for improved SRAM scalability. Solid-State Electron 65:184–190

Chapter 7
Tunnel FET (TFET)

7.1 Introduction

As discussed in the previous sections, the SS of the MOSFET governed by the Boltzmann tyranny (herein, the theoretical limit of SS is ~ 60 mV/decade at 300 K) is a main bottleneck in scaling down the power supply voltage (V_{DD}) as well as extensively reducing the power consumption in integrated circuits (ICs). As the gate voltage lowers the height of the channel potential barrier, the electrons in the source region move into channel region by the thermionic emission process [1]. The drain current (I_D) of the MOSFET increases exponentially (i.e., linearly increased in the semi-log plot of I_D vs. V_G, as shown in Fig. 7.1) because the electrons in the source region are distributed exponentially in the conduction band (by Boltzmann distribution). Therefore, even if the gate voltage controls the channel potential in a perfect manner, 60 mV/decade is the best value that can be obtained for SS at room temperature. Thus, a new transistor with a sub-60-mV/decade SS plays a critical role in the future low-power applications.

One of the most popular alternatives for implementing the sub-60-mV/decade SS is the Tunnel FET (TFET). The asymmetric structure of the TFET is the conspicuous and distinct difference from the MOSFET (i.e., p-i-n for source-channel-drain in n-type TFET vs. n-i-n for source-channel-drain in n-type MOSFET), as shown in Fig. 7.2a [2, 3]. This asymmetric device structure is associated with the band-to-band carrier injection mechanism, resulting in steeper turn-on features for the TFET. Figure 7.2b shows the operating principle of an n-type TFET. In the MOSFET, electrons in the conduction band of the source region are injected over the potential barrier. However, in the TFET, the electrons in the valence band of the source region are injected into the channel region when the conduction band of the channel region is aligned with the valence band of the source region. This carrier injection process is called band-to-band tunneling (BTBT) [4–6]. The BTBT current can be expressed as follows [6]:

© Springer Science+Business Media Dordrecht 2016
C. Shin, *Variation-Aware Advanced CMOS Devices and SRAM*,
Springer Series in Advanced Microelectronics 56,
DOI 10.1007/978-94-017-7597-7_7

Fig. 7.1 Drain current (I_D) versus gate voltage (V_G) of MOSFET and TFET

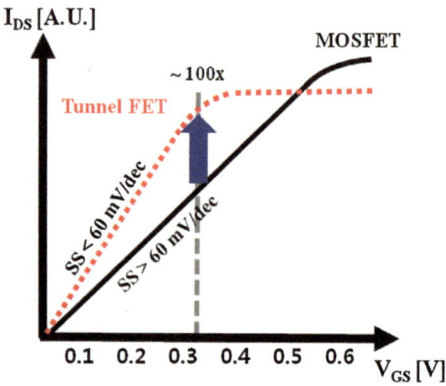

Fig. 7.2 a Illustrated cross-sectional view and **b** energy band diagram of n-type TFET

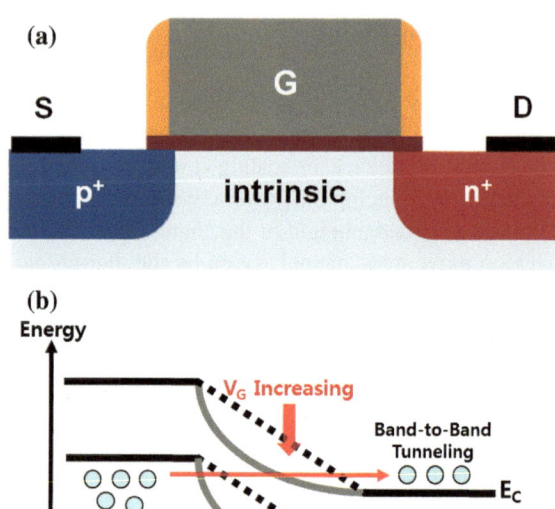

$$I_{BTBT} = A\varepsilon^2 e^{-\frac{B}{\varepsilon}} \tag{7.1}$$

where A and B are material parameters and ε is the electric field. The electron distribution does not limit the SS because the energy band gap (E_G) cuts off the Boltzmann "tail" of the electrons in the p-type source region of the TFET. With a lesser dependence on thermal voltage (i.e., kT/q), a steeper SS (i.e., sub-60 mV/decade SS) can be achieved in the TFET (see Fig. 7.1), and it is experimentally reported in [7]. As shown in Fig. 7.1, the overdrive voltage (i.e., $V_G - V_T$) can be maintained in the TFET at a much lower power supply voltage because of the steeper SS. In this regard, the TFET is a promising steep-switching device for low-power analog circuit applications.

However, the on-state drive current of the TFET is quite low compared to the conventional MOSFET, because the region where BTBT occurs is narrow and limited in the TFET. In order to improve the on-state drive current, a heterogeneous junction structure is used in the TFET rather than a homogeneous junction structure. Because the B parameter in exponential term in (7.1) is decreased with a lower energy band gap and lighter tunneling mass, the use of materials with lower energy band gaps and lighter tunneling masses, such as germanium (Ge, $E_G = 0.66$ eV) [8] and indium arsenide (InAs, $E_G = 0.36$ eV) [9], can exponentially increase the BTBT current [10]. Furthermore, a gate-to-source overlap structure or a raised-source structure is proposed [10, 11], which increases the BTBT current directly by increasing the area where BTBT occurs.

7.2   Random Dopant Fluctuation (RDF) in TFET

As TFET is suggested as a replacement device in low-power applications, the process-induced random variations in the TFET should be rigorously controlled in order to operate the TFET at ultra-low power supply voltages. In this context, the RDF-induced performance variations in the TFET were investigated with technology computer-aided design (TCAD) tools. It should be noted that the Sano's model, which is generally used in commercial TCAD tools to create randomly distributed doping profiles, considers only the long-range part of the Coulomb potential, and does not include the short-range part of it. The impact of short-range Coulombic potential on a BTBT model is not known yet. Thus, the amount of performance variation considered in this chapter would change if the short-range Coulombic potential were included. However, the results are still meaningful because they qualitatively provide the characteristics of RDF in TFETs.

TFET contains an intrinsic silicon channel region, because the TFET is relatively robust to DIBL unless the effective gate length is aggressively scaled down [12]. Thus, the randomly distributed dopants in the channel region do not have a significant influence on the RDF-induced threshold voltage variations. Moreover, the contribution of the randomly distributed dopants in the drain region to the total RDF-induced threshold voltage variation is relatively small, because the drain region (vs. source and channel regions) by itself is not closely associated with the formation of the on-state drive current. Herein, readers should note that the drain voltage in the drain region would affect the tunneling current in the TFET (the details will be mentioned later in this section). Hence, randomly placed dopants in the source region play a dominant role in determining the RDF-induced threshold voltage variations in the TFET [13]. In other words, the doping concentration of the source region directly affects the band bending at the interface of the source and channel regions, which determines the tunneling probability and the tunneling distance in the TFET. In particular, the doping profile at the source-to-channel

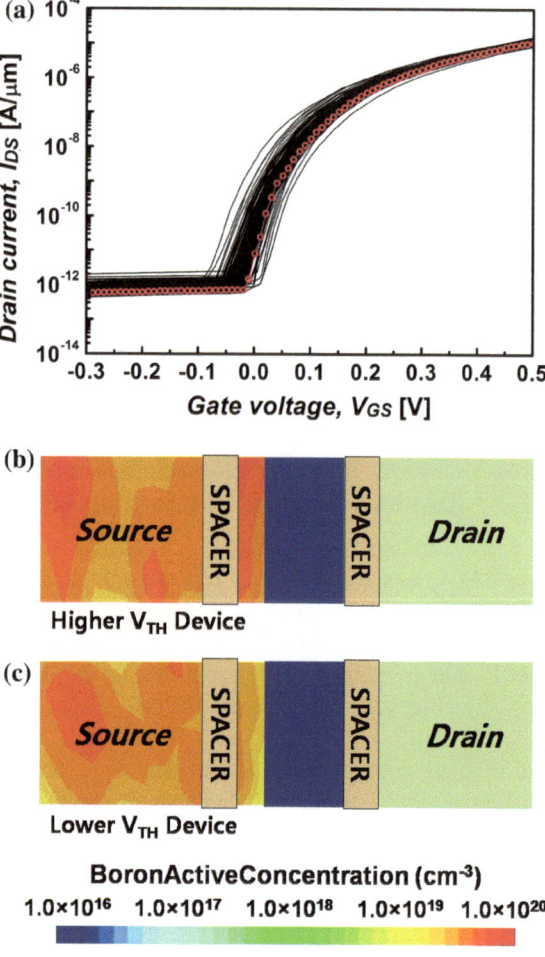

Fig. 7.3 **a** Drain current versus gate voltage (I_D vs. V_G) of overlapped Ge-source TFET [13], **b** randomized doping profile of higher threshold-voltage (V_{TH}) Ge-source TFET, and **c** randomized doping profile of lower threshold-voltage (V_{TH}) Ge-source TFET

junction or at the surface of the gate-to-source overlap region has a critical effect on the performance variations in TFETs.

Figure 7.3a shows the input characteristics of an overlapped Ge-source TFET. Figure 7.3b, c show the doping profiles of the overlapped Ge-source TFETs with a higher threshold voltage and a lower threshold voltage, respectively. Although the method of defining the threshold voltage for a TFET is still in debate [14], the threshold voltage of the TFET can be defined using the constant-current method. As shown in Fig. 7.3b, if the gate overlap region is heavily doped, the overlapped Ge-source TFET is likely to have a higher threshold voltage. On the contrary, as shown in Fig. 7.3c, if the doping concentration of the gate overlap region is relatively low, the device is likely to have a lower threshold voltage. This is because a higher gate voltage is required for the vertical BTBT generation in the heavily doped gate overlap region (because, for vertical BTBT generation, the conduction

Fig. 7.4   a I_D versus V_G for various source-doping concentrations in "raised" Ge-source TFET, **b** standard deviation of threshold voltage for various source doping concentrations [13]

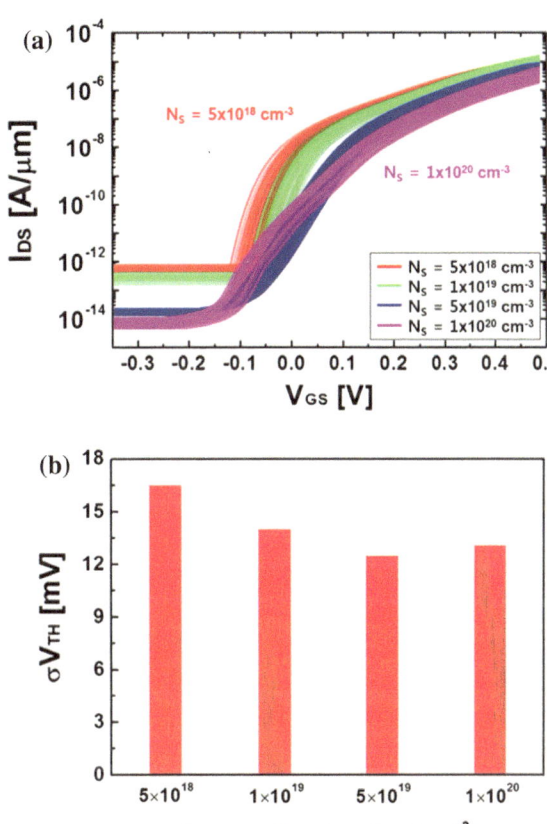

band of the gate overlap region should be lowered below the valence band). Intuitively, the probability that the dopant atoms are located at the surface of the gate overlap region is more or less low. In other words, the doping concentration at the surface of the gate overlap region is more likely to be lower than the nominal doping concentration. Therefore, the RDF in the source region tends to lower the threshold voltage of the TFET, as shown in Fig. 7.3a [13]. To further investigate the impact of RDF on the overlapped Ge-source TFET, the nominal doping concentration of the source region is varied from 5×10^{18} cm^{-3} to 1×10^{20} cm^{-3}, and the corresponding simulation results are shown in Fig. 7.4. Herein, a major difference between the lightly doped source region and the heavily doped source region exists in the dominant pathway for tunneling. In other words, the BTBT occurs within the gate overlap region in a direction perpendicular to the source/gate oxide interface in the case of lightly doped source region (i.e., vertical tunneling), whereas electrons are laterally injected/tunneled from the source region to the channel region when the source region is heavily doped (i.e., lateral tunneling).

Fig. 7.5 Drain current versus
gate voltage of raised
Ge-source TFET [17]

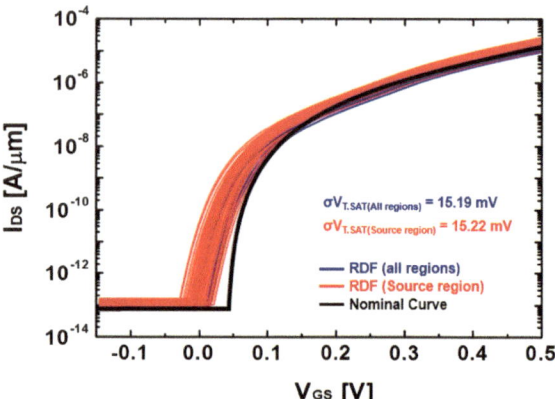

The contribution of RDF to the source region is decreased when the lateral tunneling is dominant, because the channel and source doping profiles affect the tunneling probability. Besides, the amount of RDF-induced variation is reduced because the channel doping concentration is much lower than the source doping concentration. For this reason, the overlapped Ge-source TFET with the doping concentration of 5×10^{19} cm^{-3} in its source region shows the least threshold voltage variations [13]. However, the standard deviation of the threshold voltage is slightly increased when the doping concentration is 1×10^{20} cm^{-3}, because the area where lateral BTBT is generated becomes smaller when the doping concentration is further increased [15] (this would make the TFET more sensitive to RDF).

Figure 7.5 shows the drain current versus gate voltage for the "raised" Ge-source TFETs considering the RDF in all regions, or in source region only. The input transfer characteristics of the nominal raised Ge-source TFETs are included in Fig. 7.5 to show the amount of variation between the two different cases. In the raised Ge-source TFET, the RDF tends to degrade the SS as well as decrease the threshold voltage (compared to the nominal device's threshold voltage). This originates from the fact that BTBT occurs early at the local spots (or small regions) with lower doping concentrations. Thereby, the tunneling turn-on/off voltage [16] is decreased. However, it is noteworthy that the RDF-induced threshold voltage is not significantly different in the two cases (i.e., RDF in all regions vs. RDF in the source region only) [17]. This observation indicates that, while the RDF in the channel/drain regions contributes to the total RDF-induced threshold voltage variation in the "overlapped" Ge-source TFET, the channel/drain RDF impact on the total RDF-induced threshold voltage variation is extremely reduced in the "raised" Ge-source TFET (note that the two devices have identical doping concentrations in the source, channel, and drain regions). In other words, RDF in the source region literally determines the total RDF-induced threshold voltage variation in the "raised" Ge-source TFET. This is because BTBT occurs only within the source region, and therefore, BTBT generation rate is associated only with the variation in the source doping concentration. Furthermore, the raised-source structure has a great advantage in terms of the drain-induced barrier tunneling

Fig. 7.6 Cumulative probability of DIBT for "raised" and "planar" source structures [17]

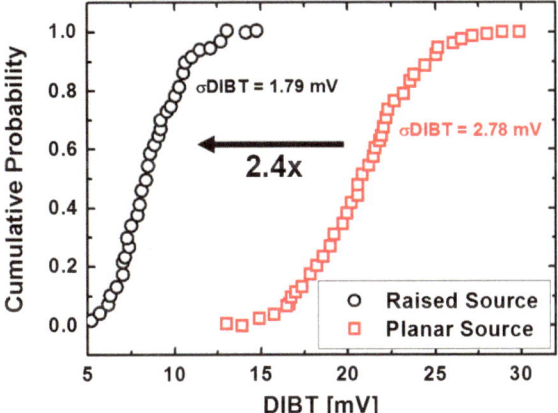

(DIBT). As shown in Fig. 7.6, even if the physical gate length of the raised Ge-source TFET is shorter than that of overlapped Ge-source TFET (e.g., 14 nm for raised Ge-source TFET [17] and 30 nm for overlapped Ge-source TFET [13]), both the DIBT value and its standard deviation are decreased significantly in the raised Ge-source TFET.

7.3 Line Edge Roughness (LER) in TFET

Because the tunneling current in the overlapped Ge-source TFET depends on the vertical BTBT generation rate at the gate-to-source overlap region, the geometry of the overlap region is as important as the doping concentration in the overlap region. Thus, provoking the geometric variation in the overlap region, the gate LER should be quantitatively characterized for satisfying the required performance specifications of the transistor. In order to understand fully how the geometry of the overlap region is associated with the LER-induced variations, simulations are conducted for two cases of source-edge profiles [18]: (1) Ge-source region and Si channel interface forms a "smooth" edge, (2) Ge-source and Si channel forms a "rough" edge, which is assumed to be perfectly correlated to the gate edge.

Figure 7.7 shows the input characteristics of an overlapped Ge-source TFET with smooth/rough LER profile. Based on the International Technology Roadmap for Semiconductors (ITRS), the LER profile for the overlapped Ge-source TFET is generated to have a root mean square roughness of 3.96 nm and a correlation length of 21.6 nm. The threshold voltage is extracted using the constant-current method. The amount of threshold voltage variation is 2.82 mV for the smooth edge LER, and 3.57 mV for the rough edge LER [18]. Considering that the number of sample devices is not statistically significant, the small difference in the threshold voltage variation can be neglected for the two cases of LER. We can say that the

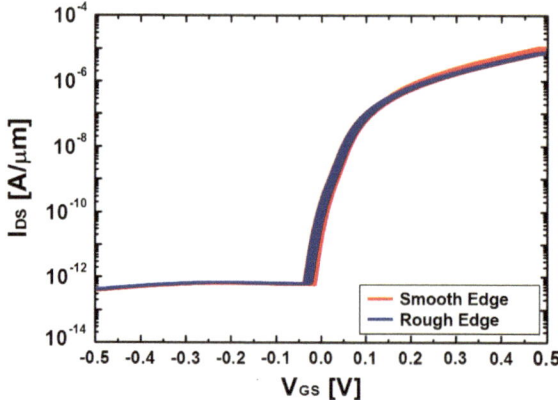

Fig. 7.7 Drain current versus gate voltage of overlapped Ge-source TFET [18]

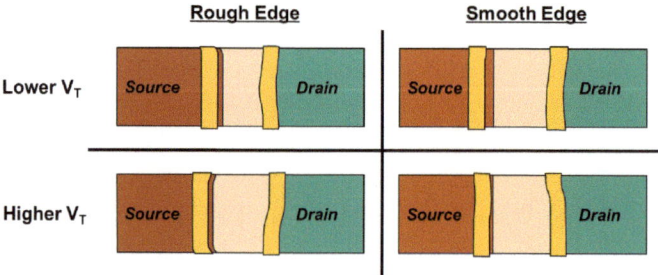

Fig. 7.8 Exemplary four different top views of rough/smooth edges with lowest/highest threshold voltages [18]

LER-induced threshold voltage variation is comparable for the two source-edge cases. In order to figure out the physical origin of the LER-induced threshold voltage variations, top views of the device structures that have the lowest (highest) threshold voltages for rough (smooth) edge cases are shown in Fig. 7.8. In case of rough edge, the variations in the effective channel length (which is defined as the shortest distance between the source and the drain [18]) cause the LER-induced threshold voltage variations. Because the electric field at the regions where tunneling occurs is affected more by the drain voltage in the short effective-channel TFET, the overlapped Ge-source TFET with a shorter effective-channel length has lesser threshold voltage variations than that with a longer effective-channel length. On the other hand, the LER-induced threshold voltage variations in the case of smooth edge are mainly caused by the area variations of the gate overlap region. Because the tunneling current depends on the tunneling area, a higher gate voltage is required to induce the current flow beyond the threshold current level if the tunneling area is decreased.

Fig. 7.9 On-state drive current versus threshold voltage (I_{ON} vs. V_T) for the two cases (i.e., smooth and rough edges) [18]

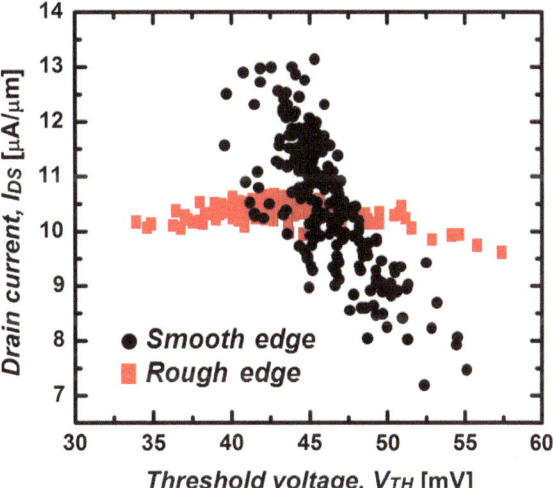

From Fig. 7.7, it can be seen that the overlapped Ge-source TFET with a smooth edge is more vulnerable to the LER-induced variations in terms of the on-state drive current (i.e., the standard deviation of the on-state drive current is 1.25×10^{-6} A/μm for the smooth-edge case, but it is 1.86×10^{-7} A/μm for the rough-edge case). The physical origin can be found using Fig. 7.8. In case of rough edges, the tunneling area is not varied much because the source edge is almost perfectly correlated with the gate LER profile. However, in the case of smooth edges, the tunneling area varies significantly from the gate LER profile. Therefore, since the on-state drive current in the overlapped Ge-source TFET is directly related to the tunneling area, the LER in the smooth-edge case leads to significant variations in the on-state drive current. Figure 7.9 shows the relationship between the on-state drive current and the threshold voltage for the two cases of source-edge profiles. A lower variation is observed in the on-state drive current with a rough source edge in the TFET, resulting from the variations in the effective-channel length. However, the variations in the effective-channel length spontaneously result in variations in the threshold voltage.

The LER-induced random variations in a double-gate lateral TFET is investigated in [19]. Two different LER profiles along the gate and channel (namely, gate LER and fin LER as mentioned in previous chapter) are considered in the double-gate device structure. However, in [19], only the fin LER is considered while running the simulations because the impact of channel-length variations caused by the gate LER on the device performance is less significant, as long as the channel length is >10 nm [20]. For instance, the amount of threshold voltage variations induced by the gate LER is much lower than that induced by the RDF [19]. However, the channel-width (not length) variations caused by the fin LER would have significant effects on the electric field at the source-to-channel tunneling interface. In order to investigate the impact of fin LER on the device performance,

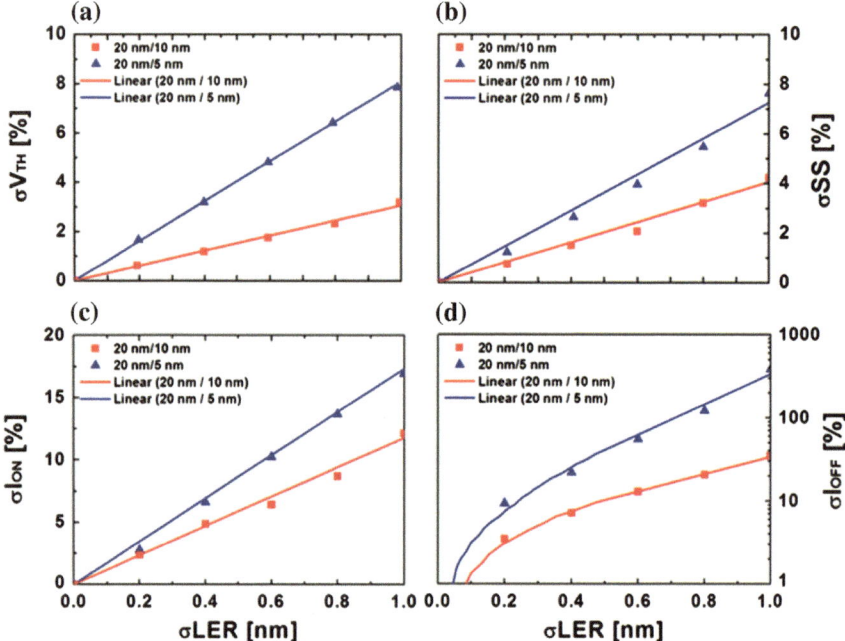

Fig. 7.10 Device performance variations [e.g., threshold voltage (V_T), on-state drive current (I_{ON}), subthreshold slope (SS), and off-state leakage current (I_{OFF})] [19]

the fin LER profile with a root mean square roughness of 1 nm and a correlation length of 15 nm is used in the simulations. Figure 7.10 shows the performance variations (e.g., threshold voltage, SS, and on/off-state currents) for two different channel widths (10 and 5 nm). For all the performance metrics, more variation is observed in a thinner TFET (i.e., TFET with a narrower channel) because more deviation (in %) is induced in smaller devices for identical LER profiles. In addition, performance variations—except for the off-state leakage current variations—increase "linearly" with increase in the root mean square roughness, while the off-state leakage current variations increase "exponentially." It is noteworthy that the magnitudes of the on-state drive current variations in the TFET are tripled compared to the inversion-mode (IM) tri-gate MOSFET with an identical device size and LER profile. This is because the channel-width variations near the interface between the source and the channel are associated with the tunneling current, whereas the on-state drive current in the IM tri-gate MOSFET is affected by the channel-width variations that are not near the source-to-channel region but along the channel length direction. Therefore, the TFET shows a higher sensitivity for an identical fin LER profile.

7.4   Work-Function Variation (WFV) in TFET

As one of the device solutions to improve the on-state drive current in the TFET, the HK/MG technique is adopted in the gate stack of the TFET because it can provide enhanced coupling between the gate voltage and the tunneling barrier (resulting in both improved on-state drive current and better SS [21]). However, the introduction of the HK/MG technology is inevitably accompanied by the WFV-induced device performance variations. From the fact that WFV is the most significant random variation source in the MOSFET, the TFET is expected to suffer from the WFV-induced random variations significantly, and thereby, the impact of WFV on TFET should be analyzed deeply. In this context, the WFV in TFET has been studied in [22–24].

First, a comparative analysis of the WFVs of TFET and MOSFET is conducted [22]. In order to compare the WFV-induced variations, TCAD simulations for TFETs and MOSFETs with TiN metal gates are carried out. TFETs and MOSFETs used in the simulations have identical device structures except for the source-doping type and the channel-doping concentrations. Figure 7.11 shows the simulated input transfer characteristics of the TFETs and MOSFETs. Performance metrics such as threshold voltage, tunneling turn-on voltage, and SS, are extracted from the input characteristic curves. Then, the amount of variation in each metric is summarized in Fig. 7.11. The amount of threshold voltage variation for TFETs is larger than that for MOSFETs because of larger SS variations in TFETs. Thus, the tunneling turn-on voltage of the TFETs should be compared to the threshold voltage of the MOSFETs, in order to exclude the influence of SS. Even without considering the SS variations, the tunneling turn-on voltage of TFETs show worse variability than the threshold voltage of MOSFETs. This is because the tunneling turn-on voltage is determined by only a few metal grains located at the narrow region near the source, whereas the impact of work function on the threshold voltage of the MOSFET is averaged out over the entire channel region. On the other hand, the SS of the MOSFET is independent of WFV, because the number of electrons injected into the channel by the thermionic emission process is determined by the carrier distribution (instead of band structure). On the contrary, the SS in the TFET is more sensitive to WFV, because the BTBT generation rate is modulated significantly by the band structure between the source and channel region. More specifically, the four extreme cases, which have the minimum/maximum values of SS and tunneling turn-on voltage, are illustrated in Fig. 7.12. It is noteworthy that the TFET has a minimum SS and tunneling turn-on voltage when the metal grains with low work function are concentrated near the source region because of steeper band bending (see Fig. 7.13).

The metric—"Ratio of average Grain size to Gate area" (RGG)—can be used to estimate the WFV-induced threshold voltage variations in the Ge-source TFETs, after the RGG has been appropriately modified for the TFET [24]. Figure 7.14 shows the amount of WFV-induced threshold voltage variation as a function of gate length/width for various average grain sizes. The WFV-induced threshold voltage

Fig. 7.11 Simulated input
transfer characteristics of
a TFETs, **b** MOSFETs.
Identical gate areas
(=56 nm × 56 nm) and
average grain sizes (=4 nm)
are used in both devices [22]

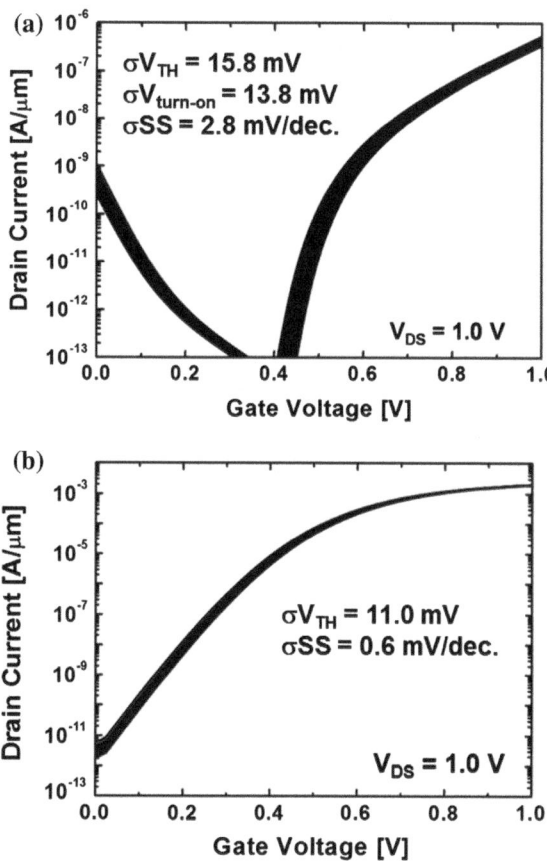

variation is suppressed with a wider gate, because the number of metal grains is increased (i.e., because of the averaging effect). However, it should be noted that the WFV-induced threshold voltage variation is not significantly high with shorter or longer gate lengths, even though the number of grains is increased. It can be explained by utilizing the device physics for TFETs (i.e., BTBT). As mentioned above, the BTBT generation rate is associated with only the band structure near the interface between the source and channel regions (i.e., within 5 nm). Therefore, even if the number of grains is increased with the gate length, the work functions of the metal grains existing far from the interface cannot affect the BTBT variations; hence, the amount of the WFV-induced threshold voltage variation is independent of the gate length. In other words, the WFV-induced threshold voltage variation is solely affected by the gate width, and thereby, the RGG concept for TFET should be modified as follows [24]:

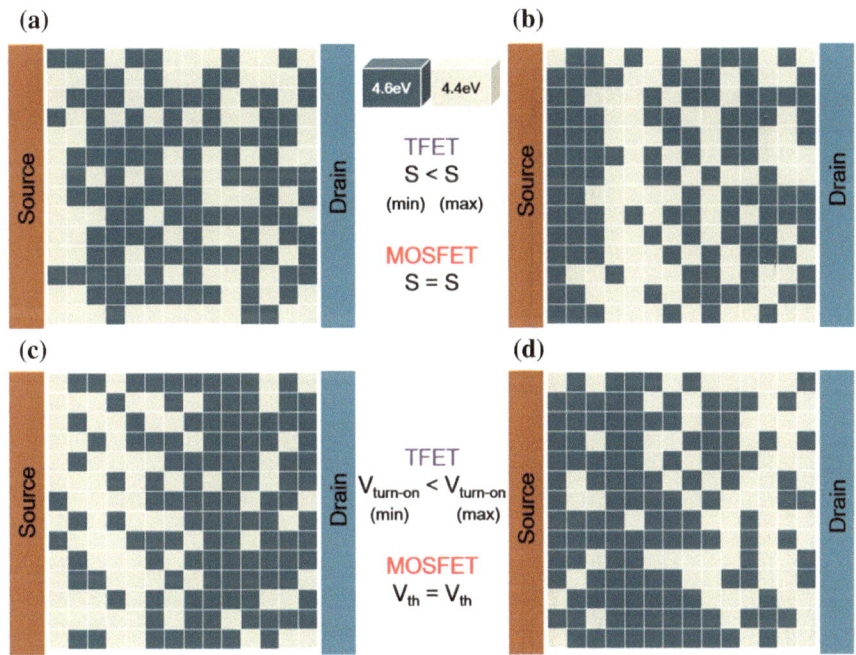

Fig. 7.12 Four extreme cases illustrating the minimum/maximum values of SS and tunneling turn-on voltage [22]

Fig. 7.13 TFET has minimum SS and tunneling turn-on voltage when metal grains with low work function are concentrated near the source region because of steeper band bending [22]

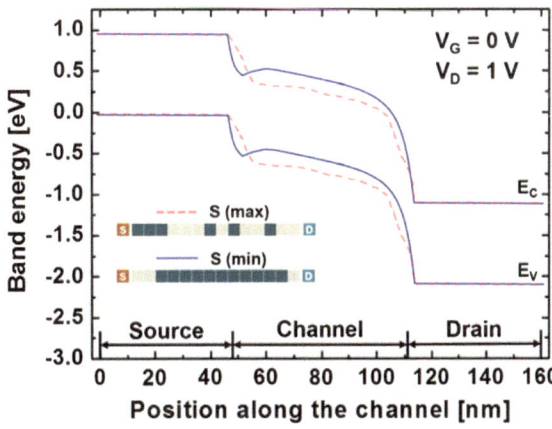

$$RGG|_{TFET} = (average\ grain\ size\ /\ gate\ width)^{0.5}$$

The standard deviation of the WFV-induced threshold voltage variations for Ge-source TFETs is plotted against the modified RGG concept in Fig. 7.15. In the RGG plot for the MOSFET, the expected WFV-induced threshold voltage

Fig. 7.14 WFV-induced
threshold voltage variation as
a function of **a** gate width and
b gate length. Note that the
average grain size is
represented as G_{size} [24]

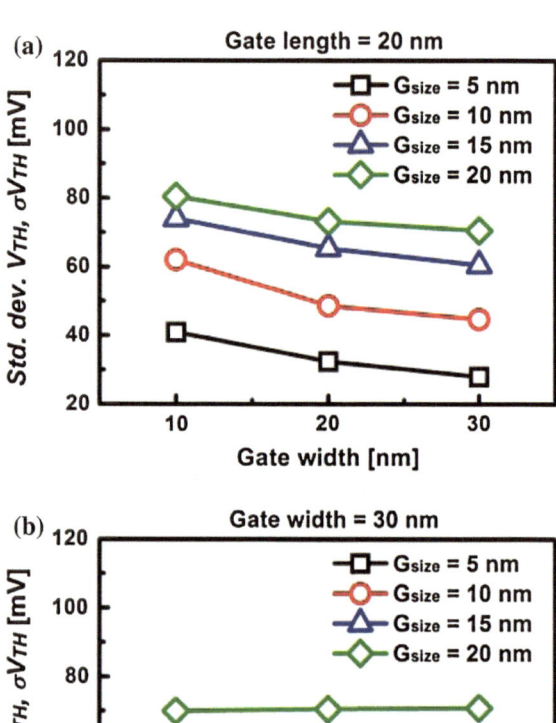

variations can be easily estimated because the slope of the RGG plot is constant for
a given metal. However, it is difficult to determine the WFV-induced threshold
voltage variations for a given RGG, because the slope of the RGG plot for
Ge-source TFETs varies depending on the average grain sizes and gate widths.
However, the slope of the RGG plot for the Ge-source TFET can be expressed as
(gate width + average grain size + 37) in the unit of mV, and the simulation results
are well matched to the analytically estimated slope (see Fig. 7.16). For further
investigation of the slope of the RGG plot for the Ge-source TFET, extensive
simulations with various oxide thicknesses, supply voltages, and tunneling
parameters (i.e., A and B parameters) are conducted, and the results are plotted in

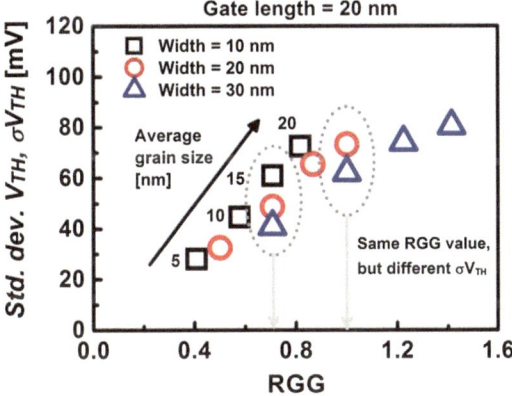

Fig. 7.15 WFV-induced threshold voltage variation versus RGG (i.e., σV_{TH} vs. *RGG*) [24]

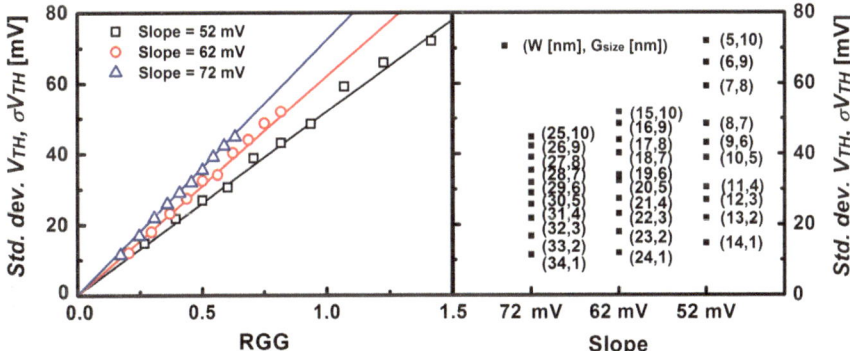

Fig. 7.16 WFV-induced threshold voltage variation versus RGG with three different slopes [24]

Fig. 7.17. An interesting fact is that the WFV-induced threshold voltage variations for the Ge-source TFET become severe as the oxide thickness is aggressively scaled down. This is because the strong gate controllability, with a thinner oxide thickness, leads to more fluctuations in the channel potential (see Fig. 7.18). However, because the drain voltages and tunneling parameters do not affect the sensitivity of the channel potential to WFV, the WFV-induced threshold voltage variation is independent of the aforementioned factors. As a result, one can reach the conclusion that the slope of the RGG plot for the Ge-source TFET is related to the gate width, the average grain size, and the oxide thickness as follows: $slope = W + G_{size} + \alpha - (\beta \times T_{ox})$ mV, where, W is the gate width, G_{size} is the average grain size, α (the value of 45) and β (the value of 16) are fitting parameters,

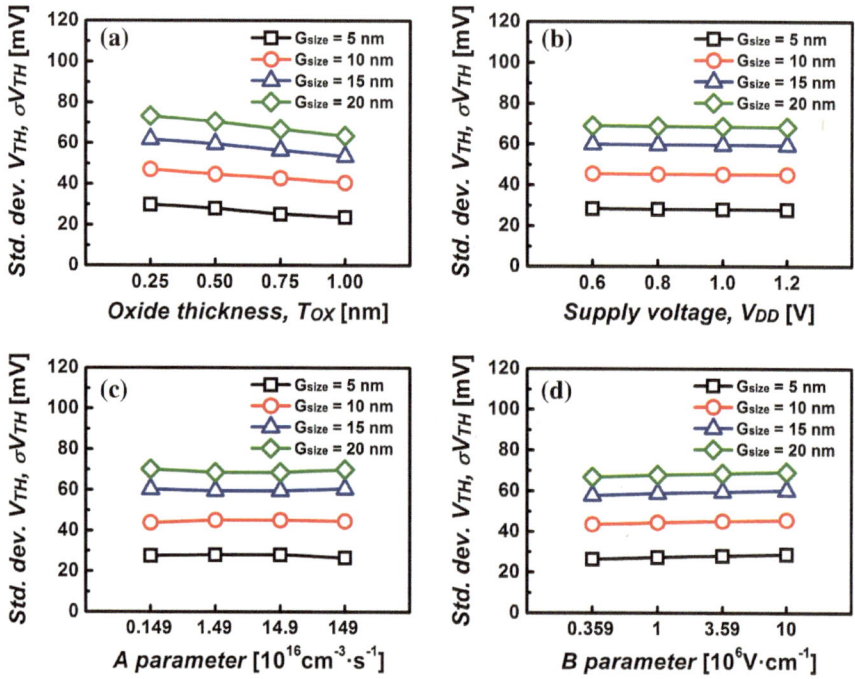

Fig. 7.17 WFV-induced threshold voltage versus **a** oxide thickness, **b** power supply voltage, **c** "*A*" parameters, and **d** "*B*" parameters, with various average grain sizes [24]

Fig. 7.18 Energy band diagram for various gate oxide thicknesses [24]

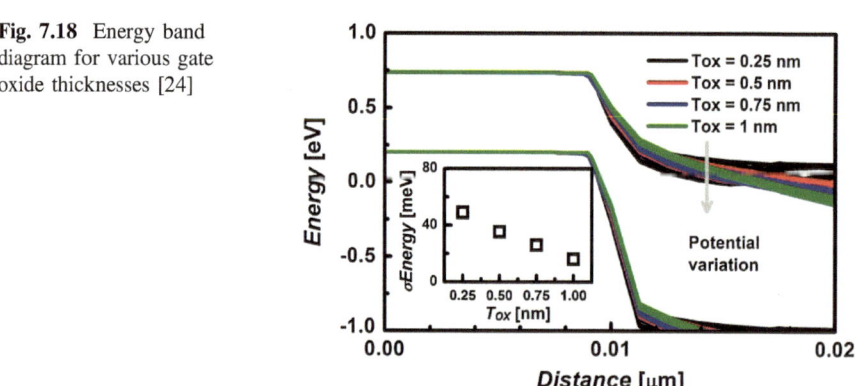

and T_{ox} is the oxide thickness. It should be noted that the slope would be used to estimate the WFV-induced threshold variations for the TFET regardless of the source material, because the tunneling parameters such as A and B parameters are irrelevant to the WFV-induced threshold voltage variations (see Fig. 7.17c and d).

References

1. Taur Y, Ning TH (2009) Fundamentals of modern VLSI devices. Cambridge University Press, New York
2. Reddick WM, Amaratunga GAJ (1995) Silicon surface tunnel transistor. Appl Phys Lett 67 (4):494–496
3. Hu CC (2010) Modern semiconductor devices for integrated circuits. Prentice Hall, Upper Saddle River, New Jersey
4. Kane EO (1959) Zener tunneling in semiconductors. J Phys Chem Solids 12(2):181–188
5. Kane EO (1961) Theory of tunneling. J Appl Phys 32(1):83–91
6. Moll JL (1964) Physics of semiconductors. McGraw-Hill, New York
7. Choi WY, Park B-G, Lee JD, Liu T-JK (2007) Tunneling field-effect transistors (TFETs) with subthreshold swing (SS) less than 60 mV/dec. IEEE Electron Devices Lett 28(8):743–745
8. Butcher PN, Hulme KF, Morgan JR (1962) Dependence of peak current density on acceptor concentration in germanium tunnel diodes. Solid-State Electron 5(5):358–360
9. Jain SC, McGregor JM, Roulston DJ (1990) Band-gap narrowing in novel III-V semiconductors. J Appl Phys 68(7):3747–3749
10. Kim SH, Kam H, Hu C, Liu T-JK (2009) Germanium-source tunnel field effect transistors with record high I_{ON}/I_{OFF}. In: Symposium on VLSI Technology Digest, pp 178–179
11. Kim SH, Agarwal S, Jacobson ZA, Matheu P, Hu C, Liu T-JK (2010) Tunnel field effect transistor with raised germanium source. IEEE Electron Device Lett 31(10):1107–1109
12. Boucart K, Ionescu AM (2007) Length scaling of the double gate tunnel FET with a high-K gate dielectric. Solid-State Electron 51(11/12):1500–1507
13. Damrongplasit N, Shin C, Kim SH, Vega RA, Liu T-JK (2011) Study of random dopant fluctuation effects in germanium-source tunnel FETs. IEEE Trans Electron Devices 58 (10):3541–3548
14. Boucart K, Ionescu AM (2008) A new definition of threshold voltage in tunnel FETs. Solid-State Electron 52(9):1318–1323
15. Vandenberghe WG, Verhulst AS, Groeseneken G, Soree B, Magnus W (2008) Analytical model for point and line tunneling in a tunnel field-effect transistor. In: Proceedings of SISPAD, pp 137–140
16. Hu C, Patel P, Bowonder A, Jeon K, Kim SH, Loh WY, Kang CY, Oh J, Majhi P, Javey A, Liu TJK, Jammy R (2010) Prospect of tunneling green transistor for 0.1 V CMOS. In: Proceedings of IEDM, pp 16.1.1–16.1.4
17. Damrongplasit N, Kim SH, Liu T-JK (2013) Study of random dopant fluctuation induced variability in the raised-Ge-source TFET. IEEE Electron Device Lett 34(2):184–186
18. Damrongplasit N, Kim SH, Shin C, Liu T-JK (2013) Impact of gate line-edge roughness (LER) versus random dopant fluctuations (RDF) on germanium-source tunnel FET performance. IEEE Trans Nanotechnol 12(6):1061–1067
19. Leung G, Chui CO (2013) Stochastic variability in silicon double-gate lateral tunnel field-effect transistors. IEEE Trans Electron Devices 60(1):84–91
20. Bhuwalka KK, Sedlmaier S, Ludsteck AK, Tolksdorf C, Schulze J, Eisele I (2004) Vertical tunnel field-effect transistor. IEEE Trans Electron Devices 51(2):279–282
21. Boucart K, Ionescu AM (2007) Double-gate tunnel FET with high-κ gate dielectric. IEEE Trans Electron Devices 54(7):1725–1733
22. Choi KM, Choi WY (2013) Work-function variation effects of tunneling field-effect transistors (TFETs). IEEE Electron Device Lett 34(8):942–944
23. Choi KM, Lee W-S, Lee K-H, Park Y-K, Choi WY (2015) Influence of preferred gate Metal grain orientation on tunneling FETs. IEEE Trans Electron Devices 62(4):1353–1356
24. Lee Y, Nam H, Park J-D, Shin C (2015) Study of work-function variation for high-κ/metal-gate Ge-source tunnel field-effect transistors. IEEE Trans Electron Devices 62 (7):2143–2147

Part III
Static Random Access Memory (SRAM) Based on Advanced CMOS Devices

Chapter 8
Applications in Static Random Access Memory (SRAM)

8.1 Introduction

Continuous efforts to shrink the physical size of transistors enable the integration of a larger number of transistors on a single chip. Today, it is feasible to fabricate a single SRAM cell in a 0.04999-μm^2 area [1]. However, as the transistors are aggressively scaled down, the impacts of the process-induced random variations on the device performance are dramatically increased [2]. The term "process-induced random variations" indicates that the process/fabrication steps such as doping, lithography, etching, chemical mechanical polishing (CMP), and so on, induce these random variations [3]. The well-known random variation sources caused by the aforementioned process steps are random dopant fluctuations (RDF) [4] [also known as random dopant distributions (RDD)] [5], line edge roughness (LER) [6], work function variations (WFV) [7] (also known as metal grain granularity (MGG) [8]), and oxide thickness variations [9]. Aforementioned random variations cause variations in parameters such as effective channel length and threshold voltage of each transistor. Then, the mismatches in the effective channel lengths and threshold voltages between two neighboring transistors cause degradations in the stability of the SRAM cells, which consist of a few transistors (e.g., 6 transistors for 6T SRAM cell) [10]. This is because each transistor of a single SRAM cell should be balanced in order to make it stable for the read/write/retention operations [11]. Further detailed information about how to balance the different transistors of a single SRAM cell, and how badly the stabilities of the SRAM cells are degraded by process-induced variations, will be provided later.

Since random variations significantly affect the performance of SRAM cells, it is irrefutable that random variations are one of the key factors in the SRAM cell design. Furthermore, it is important to look for "random variation"-tolerant SRAM designs because random variations cannot be controlled, as systematic variations are controlled [12]. In order to find the random-variation–tolerant SRAM cell designs, understanding how the SRAM cell operates and how the process-induced

© Springer Science+Business Media Dordrecht 2016   123
C. Shin, *Variation-Aware Advanced CMOS Devices and SRAM*,
Springer Series in Advanced Microelectronics 56,
DOI 10.1007/978-94-017-7597-7_8

random variations affect the SRAM cell operations, is important. Furthermore, it is necessary to understand how the SRAM cell layout affects both the stability and the random-variation tolerance of the SRAM cells, because the properties of the SRAM cells are determined by not only the electrostatic properties of the transistors but also the SRAM cell layout designs/styles [13]. Then, the understanding of the SRAM cell operations and the SRAM cell layouts become the foundation for understanding the SRAM margin metrics (which are the quantitative indexes representing the stability of the SRAM cell). Hence, the operating principle of the SRAM (i.e. read and write operations of SRAM) will be explained in Sect. 8.2. Then, we will go over the SRAM cell layout design in Sect. 8.3. In Sect. 8.4, we will review the SRAM margin metrics (especially, the conventional SRAM margin metrics and the large-scale SRAM margin metrics). Lastly, we will discuss how to perform the simulations to estimate the yield of the SRAM array. However, the reader may require fundamental knowledge about statistics to understand Sect. 8.4 in a better manner.

8.2   The Operating Principle of SRAM

There are many types of SRAM bit cells; however, we focus only on the operation of a six-transistor (6T) SRAM cell in this section, since the 6T SRAM cell is the most popular SRAM cell because of its superior robustness and low-power operations [14] along with high capacity.

As shown in Fig. 8.1, the 6T SRAM cell consists of four n-type MOS (NMOS) transistors and two p-type MOS (PMOS) transistors. At a first glance, we can figure out that the PMOS and NMOS, which are connected in series, compose a CMOS inverter (note that a CMOS inverter circuit is shown in Fig. 8.2a). With that in mind, we can illustrate a 6T SRAM cell, as shown in Fig. 8.2b. It is more convenient to understand the operating principle of the SRAM cell with the help of

Fig. 8.1 Circuit schematic of 6T SRAM bit cell

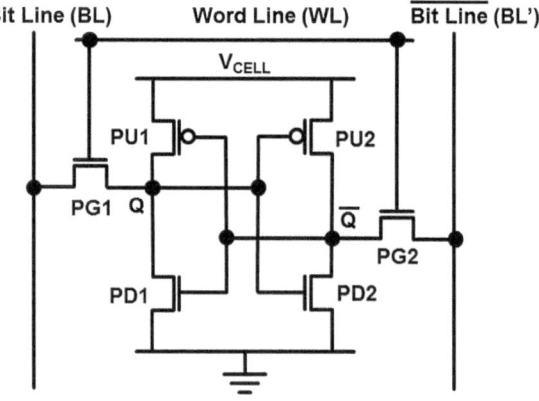

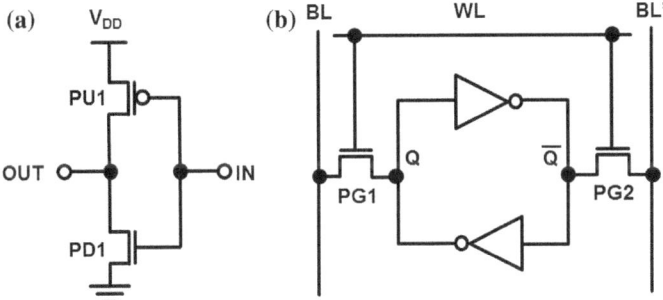

Fig. 8.2 **a** CMOS inverter circuit and **b** 6T SRAM bit cell with cross-coupled inverters

Fig. 8.2b. Data is stored in the cross-coupled inverter latch (the output of an inverter is connected to the input of another inverter and vice versa), and the data can be accessed and modified through two NMOS transistors (which are connected to the bit lines and the cross-coupled inverter latch).

For analyzing the operation of the SRAM bit cell in more detail, the basic operations of the SRAM cell can be divided into three: The first operation is the "read operation." In the read operation, the data stored in the SRAM bit cell is accessed through the bit lines. However, the data stored in the SRAM bit cell must not be disturbed (i.e., the data stored in SRAM bit cell cannot be changed from 0 to 1 during the read operation). If this data is disturbed during the read operation, it is called a destructive read operation [15]. The second operation is the "write operation." In the write operation, the data stored in the SRAM bit cell is flipped (or changed) from the previous value to a new value (i.e., from 0 to 1, or from 1 to 0). In this operation, it is important to make the data-flipping process fast and reliable. The last operation is the "data retention operation." Since the bit lines (i.e., BL and BL') are disconnected from the SRAM bit cell by two pass-gate transistors (i.e., PG1 and PG2), the SRAM bit cell works just like as a cross-coupled inverter latch.

8.2.1 Read Operation

Both the read and write operations of the 6T SRAM cell are similar in some aspects. First, the BL and BL' are pre-charged [16]. Then, the pass gates are turned on by applying voltages to the gate electrodes of the pass-gate transistors through the word line (WL). Note that the cross-coupled inverter latch is always powered through the V_{cell} line (where, usually, V_{DD} is biased). The key difference between the read operation and the write operation is in the biasing conditions on the bit lines.

To access the data stored in the cross-coupled inverter latch without flipping the stored data, both the bit lines (i.e., BL and BL') of a 6T SRAM cell are pre-charged to V_{DD} (or V_{HI}, which is the voltage level considered as the high state or bit '1').

Note that the parasitic capacitance at the bit lines in the SRAM circuit is neglected. Generally, the bit line parasitic capacitance can be divided into three types of parasitic capacitances: (i) metal-wire capacitance of the bit lines, (ii) gate-to-drain overlap capacitance, and (iii) junction capacitance of the source and drain of the pass-gate transistors [17]. The metal-wire capacitance of the bit lines is a combination of (1) the capacitances between the BL and BL' metal wires, (2) the capacitance between the bit line and the other bit lines of neighboring SRAM cells, and (3) the capacitance between the bit line and the ground line (or *GND*, in short) [18]. The metal-wire capacitance of the bit line exists because there is always a capacitance between two metal wires or plates. The root-cause of the junction capacitance is quite different from that of the metal-wire capacitance. The cause of the junction capacitance between the source and drain of the pass-gate transistors is the depletion region surrounding the source and drain of the pass gate transistors.

After the bit lines are pre-charged to V_{DD}, the next step in the read operation of the 6T SRAM cell is turning the pass gates on, by applying a voltage through the word line. Consider the case where, the bit '0' is stored in a 6T SRAM bit cell. Then, the voltage at nodes Q and $\bar{Q}$ is V_{DD} and *GND*, respectively. Right after the bit lines are pre-charged and the pass gates are turned on, the voltage at each node will be as shown in Fig. 8.3. Since the voltage at the node Q is *GND*, the PD2 is turned off. Similarly, the PU1 is turned off because the voltage at node $\bar{Q}$ is V_{DD}. Then, we can figure out that the current will flow through BL–PG1–PD1, and thereby, the pre-charged BL will be discharged (see the green line in Fig. 8.3). At last, the voltage difference between BL and BL' is detected by a sense amplifier located between the bit lines. As a result, we can read the 6T SRAM bit cell data as '0' because the voltage at BL is discharged while BL' remains unchanged. Note that there is no current flow through PG2 while PG2 is turned on. This is because the voltage difference between the drain and the source of PG2 is zero (see Fig. 8.3).

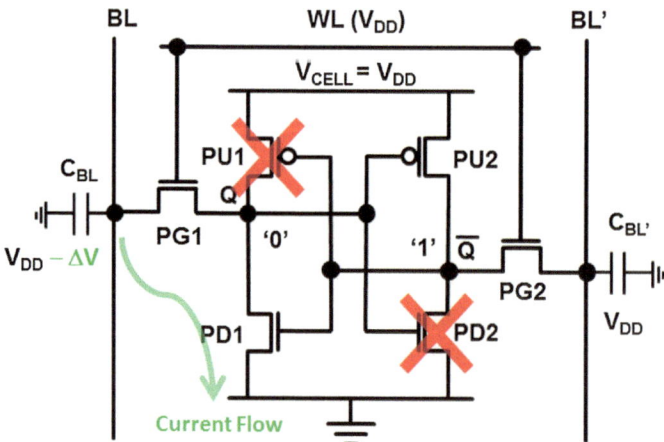

Fig. 8.3 Current flow and voltage level at each node during the read operation. The initial data stored in the SRAM cell is '0' (i.e., Q = 0)

The data stored in a 6T SRAM bit cell is detected by discharging the BL. As a result, the voltage at the node Q can be slightly increased because of the current flow through BL–PG1–PD1. If the voltage at the node Q becomes high enough to turn on PD2, and turn off PU2, the data stored in the 6T SRAM bit cell can be flipped. Therefore, PD1 must be stronger than PG1. In other words, PD1 must flow much more current than PG1 under the biasing condition for the read operation. This means that the stability of the read operation is a battle between the pass gate and the pull-down gate in the 6T SRAM bit cell. Therefore, a ratio can represent the read stability of the 6T SRAM cells. This ratio is defined as PD : PG and is named beta (β) ratio (also called cell ratio or CR) [19]. If the mobility of both PD and PG is identical, the β ratio can be simplified to the ratio of the physical dimensions (i.e., W_{ch}/L_{ch}) of PD to that of PG.

The read operation in the case in which the data stored in a 6T SRAM cell is '1' at the node Q is shown in Fig. 8.4. It is almost identical to the case when the stored data is '0' at the node Q. By just flipping the Fig. 8.3 horizontally, readers can know that it is identical to Fig. 8.4. Therefore, the read operation in the case where the data stored in a 6T SRAM cell is '1' at the node Q can be explained in the same manner as the read operation in the case that the data stored in a 6T SRAM cell at the node Q is '0'.

The channel width of the planar bulk CMOS transistors can be controlled without many limitations. However, the channel width of the FinFET (fin FET) cannot be controlled as conveniently as that of the planar bulk CMOS transistors. The fin width and height is limited by the aspect ratio, which must be followed [20] to retain its superior electrostatic properties and robustness over the planar devices. In addition, the channel width of the FinFET is quantized. Hence, only by the number of fins can modify the channel width of the transistors in a FinFET-based 6T SRAM cell.

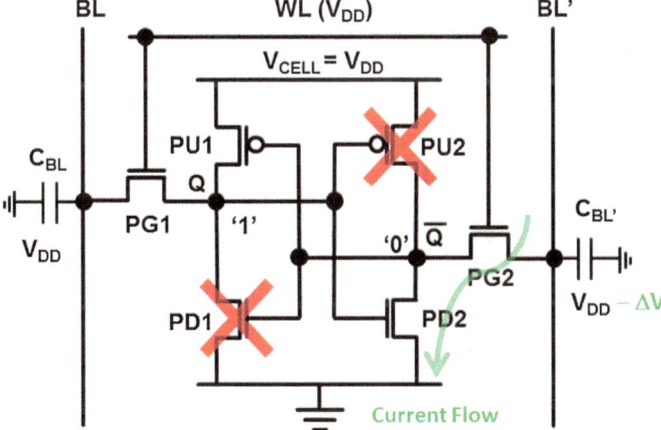

Fig. 8.4 Current flow and voltage level at each node during the read operation. The initial data stored in SRAM cell is '1' (i.e., Q = 1)

8.2.2 Write Operation

The write operation can be divided into four cases as follows—Case 1: when '0' is stored and writing '0' is requested, Case 2: when '0' is stored and writing '1' is requested, Case 3: when '1' is stored and writing '0' is requested, and Case 4: when '1' is stored and writing '1' is requested. Cases 1 and 4 are the trivial cases, which write the same data over the stored data. Therefore, Cases 2 and 3 will be discussed in this section.

To read the data in a 6T SRAM bit cell, both the BL and BL' are pre-charged to V_{DD}. However, to write the data into the SRAM cell, the pre-charged voltage levels on the BL and BL' must be opposite to each other. In order to write '0' at the node Q in the SRAM cell, the BL is driven from V_{DD} to GND while BL' is pre-charged to V_{DD}. In contrast, in order to write '1' at the node Q in the SRAM cell, BL is pre-charged to V_{DD} and BL' is driven from V_{DD} to GND. After the BL and BL' are set to the desired voltage levels, both PG1 and PG2 are turned on by applying V_{DD} to their gate electrodes via the word line. Afterwards, the voltage at each data storage node in the SRAM cell will be as shown in Figs. 8.5 and 8.6 for Case 2 and Case 3, respectively.

In Case 2 (i.e., when '0' is stored and writing '1' is requested.), after PG1 and PG2 are turned on, the current flows through BL–PG1–PD1 and PU2–PG2–BL' (see Fig. 8.5). As a result, the voltage at the node $\bar{Q}$ drops, and the voltage at the node Q increases until (i) the voltage at the node $\bar{Q}$ is low enough to turn PU1 on and PD1 off and/or (ii) the voltage at the node Q is high enough to turn PU2 off and PD2 on. Then, we can say that, the voltage levels at the nodes: Q and $\bar{Q}$, are flipped to V_{DD} and GND, respectively.

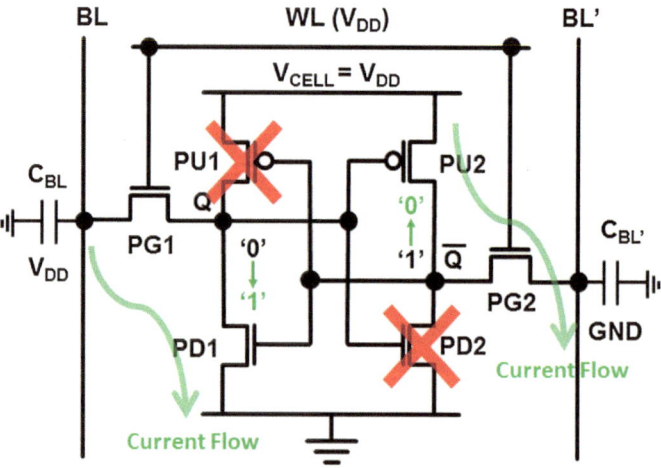

Fig. 8.5 Current flow and voltage level at each data storage node (i.e., Q and $\bar{Q}$) when writing '1' at the node Q is required. The initial data stored at the node Q was '0'

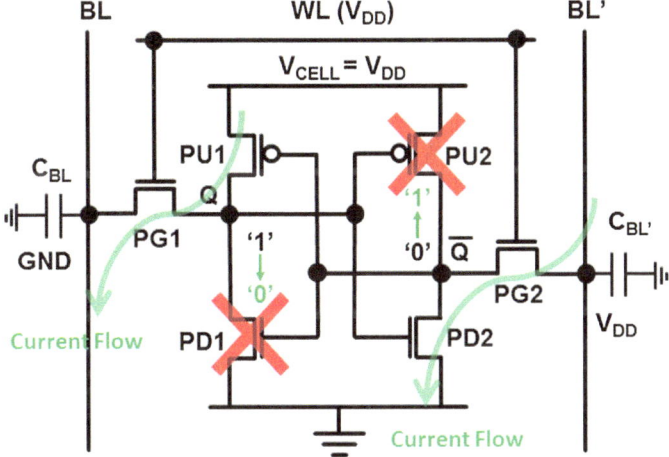

Fig. 8.6 Current flow and voltage level at each data storage node (i.e., Q and Q̄) when writing '0' at the node Q is required. The initial data stored at the node Q was '1'

Case 3 (i.e., when '1' is stored and writing '0' is requested.) can be understood in the same manner as Case 2. Since BL' is pre-charged to V_{DD} and BL is driven from V_{DD} to GND, the current flows through PU1–PG1–BL and BL'–PG2–PD2 after PG1 and PG2 are turned on (see Fig. 8.6). As a result, the voltage at the node Q drops, and the voltage at the node Q̄ increases until (i) the voltage at the node Q is low enough to turn PU2 on and PD2 off and/or (ii) the voltage at the node Q̄ is high enough to turn PU1 off and PD1 on. Then, we can say that the voltage levels at the nodes Q̄ and Q are flipped to V_{DD} and GND, respectively.

Based on Cases 2 and 3, it can be observed that PG must be stronger than PU to make sure that the data in the SRAM cell is flipped. In other words, PG must be stronger than the feedback inverter consisting of PU and PD. Hence, the gamma

Fig. 8.7 An exemplary plot showing the trade-off between the read and write noise margins. Herein, WRRV and WWTV are the metrics for read and write noise margin, respectively (see Sect. 8.3.2 for more details)

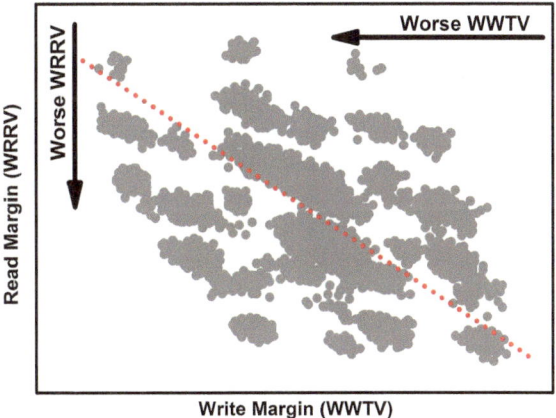

ratio (which is defined as PG : PU) represents the writeability of the 6T SRAM cell. It is the inverse of the pull-up ratio (or PR) of the SRAM cell [21]. However, there is a trade-off between the writeability and the read stability of the SRAM cell. As shown in Fig. 8.7, the writeability and read stability cannot be improved together. Therefore, it is important to find an optimal condition for balancing the read stability and writeability. This is one of the main reasons behind performing the SRAM yield estimation in terms of the read stability and writeability.

8.3 Metrics for SRAM Read/Write Noise Margin

A few qualitative factors that represent the read stability and writeability of 6T SRAM cells have already been introduced in the previous sections. However, those factors cannot be used for quantitatively determining whether the read/write operations are going to fail or not. To estimate the yield of the SRAM cells, we need quantitative metrics to predict the read and write failures of the 6T SRAM cells. The metrics designed for this purpose are the SRAM read/write noise margins.

There are two kinds of SRAM margin metrics: one for predicting the read operation failures and the other for predicting the write operation failures. The first one is called the SRAM read margin metric, and the second one is called the SRAM write margin metric. These margin metrics enable quantitative estimation of the noise margin of the SRAM cell, when the SRAM cell does not cause a failure in the read/write operation.

A few conventional metrics for the SRAM read/write noise margin are the static noise margin (SNM), the static current noise margin (SINM), and the static voltage noise margin (SVNM) for the SRAM read stability, and the write noise margin (WNM) and the writeability current (I_W) for the SRAM writeability [21]. However, these conventional SRAM metrics have weaknesses at some points; the major weakness is that additional contacts/pads are required at the storage nodes of Q and $\bar{Q}$ in the SRAM cell, to measure the conventional SRAM margin metrics directly [22].

To overcome the limitations of using the conventional SRAM margin metrics, large-scale SRAM margin metrics were proposed in 2009 [22]. The large-scale SRAM margin metrics can be directly measured through the bit lines (i.e., without using the padded-out SRAM cells). Therefore, the large-scale SRAM margin metrics are adequate for the state-of-art large-scale SRAM arrays (e.g., 64 Mb SRAM arrays [23], which include 64×2^{20} cells on a single chip). In this chapter, we will study the conventional SRAM margin metrics and the large-scale SRAM margin metrics; the conventional SRAM margin metrics are still the most widely used SRAM margin metrics, and are well associated with the large-scale SRAM margin metrics.

8.3.1   Conventional Metrics for SRAM Read/Write Operation

The read static noise margin (RSNM) of an SRAM cell can be defined using the voltage transfer characteristic (VTC) curves of the cross-coupled inverters in the SRAM cell. The butterfly curve (which is shown in Fig. 8.8b) consists of two VTC curves (i.e., one for the voltage measured at the node Q and the other for the voltage measured at the node $\bar{Q}$ of SRAM cell) during the read operation. To avoid the underestimation or overestimation of RSNM, the largest square that can be drawn within the two VTC curves of the cross-coupled inverters is defined as the RSNM. This definition was first proposed by Hill [24] in 1968. The RSNM can be easily determined in a "45° rotated" Cartesian coordinate system: In the new coordinate system, the difference between the two inverter VTC curves is calculated. Then, the magnitude between the maxima and minima points of the VTC curves is found to be the RSNM $\times \sqrt{2}$ (see Fig. 8.8b). The biasing conditions for measuring the RSNM are illustrated in Fig. 8.8a. Herein, note that both BL and BL' are pre-charged to V_{DD}. Then, the voltage at the node Q (or $\bar{Q}$) is swept from 0 to V_{DD} while measuring the voltage at the node $\bar{Q}$ (or Q). Because of its simplicity, the RSNM is the most common metric used for the SRAM read stability. If there is no square between the VTC curves of the SRAM cell, it implicitly means that the data stored in the SRAM cell would be flipped during the read operation. There is another way to define the read stability of the SRAM cells. It uses two metrics called the static voltage noise margin (SVNM) and the static current noise margin (SINM) [22]. These metrics are defined from the N-curve [25]. The N-curve (see Fig. 8.9b) is a current versus voltage curve, while the VTC is a voltage versus voltage curve. The N-curve is measured by sweeping the voltage at Q (or $\bar{Q}$) and measuring the current flow into the $\bar{Q}$ or Q node as shown in Fig. 8.9a. Then, the SVNM is determined as the minimum voltage difference between the first two points where the N-curves cross the line: $I = 0$. SINM is defined as the peak value

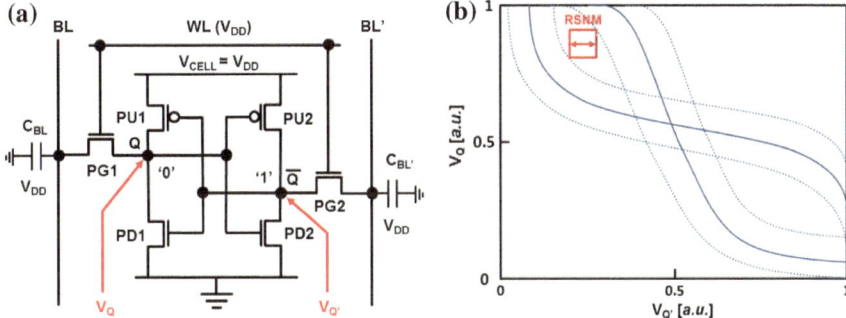

Fig. 8.8 **a** Biasing condition to measure the read static noise margin (RSNM) of the 6T SRAM cell and **b** voltage transfer characteristic (VTC) curves to measure the RSNM

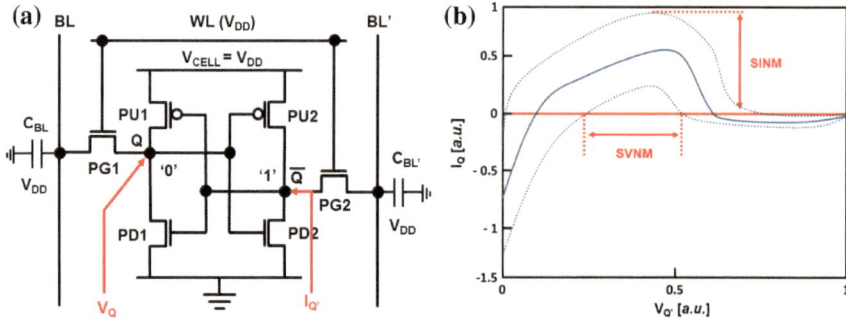

Fig. 8.9 a Biasing condition to measure static current/voltage noise margin and **b** N-curve to measure static current noise margin (SINM) and static voltage noise margin (SVNM)

of the current in the N-curves. SVNM refers to the maximum voltage noise margin as long as the data is not disrupted during the read operation. Likewise, SINM refers to the maximum current noise margin as long as the destructive read operation does not occur.

The SINM and the SVNM from the N-curve (see Fig. 8.9b) can provide information about the current noise margin of the SRAM cell, which cannot be obtained from the SNM using the VTC curves. Hence, SINM and SVNM will be useful if information about the current noise margin is required. The correspondence/association between SNM and SINM can be verified from the fact that the three points where the N-curves cross the line: "I = 0" in Fig. 8.9 correspond to the three points where the two VTC curves cross each other in Fig. 8.8.

Like the read stability of the SRAM cell, the writeability of the SRAM cell can also be defined from the VTC curves and the N-curves. The write noise margin (WNM) is determined from the VTC curves shown in Fig. 8.10b. The VTC in Fig. 8.10b is measured under the biasing condition shown in Fig. 8.10a. It is noteworthy that the voltage pre-charged (or driven) at BL and BL' is identical to the

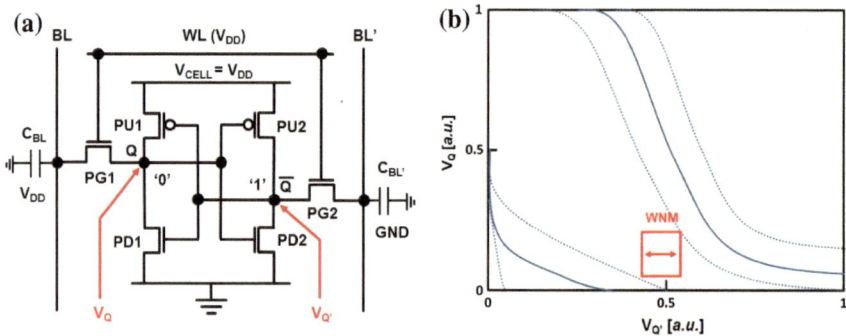

Fig. 8.10 a Biasing condition to measure write noise margin (WNM) and **b** VTC curves to measure WNM

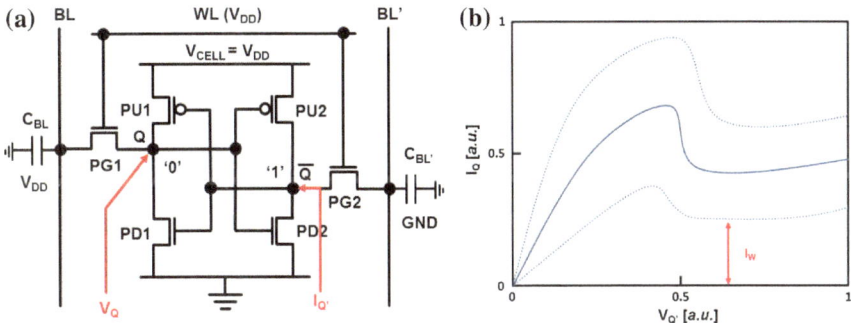

Fig. 8.11 a Biasing condition to measure writeability current (I_W) and **b** N-curve to measure I_W

biasing condition for the write operation of the SRAM cell, to measure one of two VTCs. The other VTC curve is measured under a biasing condition, identical to the one used to measure the VTC for RSNM. From the VTCs shown in Fig. 8.10b, WNM is defined as the smallest square that can be embedded between the read VTC curve and the write VTC curve. If the two VTC curves cross (and thereby WNM becomes zero), it means that the SRAM cell cannot be written into correctly.

The writeability current (which represents the current noise margin during the write operation) is defined from the N-curves (see Fig. 8.11b). Unlike the N-curves used for SVNM and SINM, these N-curves are measured when BL (or BL') is pre-charged to V_{DD} while BL' (or BL) is driven to *GND* (see Fig. 8.11a). As shown in Fig. 8.11b, I_W is defined as the minimum current in the region where V_Q (or $V_{\bar{Q}}$) is higher than the trip point of the inverter in the SRAM cell. If I_W is lesser than zero, it means that the write operation cannot be completed in the SRAM cell.

8.3.2   Large-Scale Metrics for SRAM Read/Write Operation

Unlike the conventional metrics for the SRAM read/write operations, all of the large-scale metrics are defined from the "current versus voltage" curves. To measure the supply read retention voltage (SRRV), both BL and BL' should be pre-charged to V_{DD}. Then, while V_{cell} is swept from V_{DD} to 0, the current flow through BL is measured (see Fig. 8.12a). As a result, we can obtain the transfer curves shown in Fig. 8.12b. From the curves in Fig. 8.12b, SRRV is defined as the voltage difference between V_{DD} and the maximum V_{cell} point where the measured current drops abruptly. If the SRRV is greater than zero, it means that V_{cell} can be dropped below V_{DD} without disturbing the data during the read operation.

The word line read retention voltage (WRRV) is defined from the measured current versus voltage curves (see Fig. 8.13a). The difference between measuring the WRRV and the SRRV is that the word line voltage is increased above V_{DD} for WRRV. The WRRV is defined as shown in Fig. 8.13b. If the WRRV is greater than

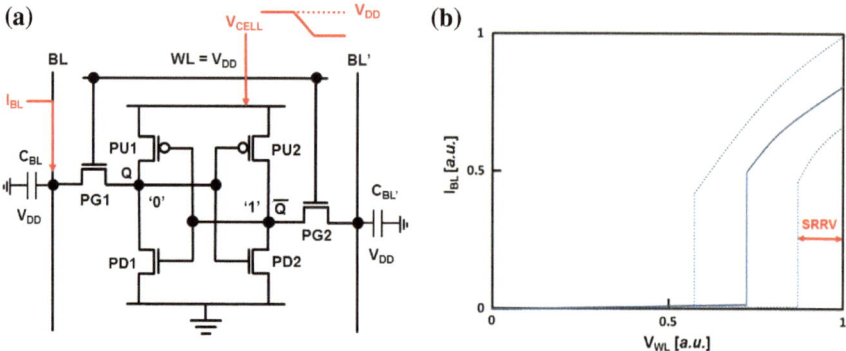

Fig. 8.12 a Biasing condition to measure supply read retention voltage (SRRV) and **b** transfer curves to measure SRRV

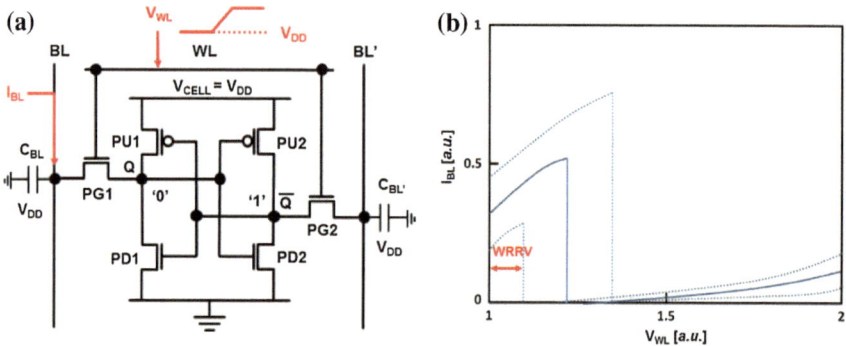

Fig. 8.13 a Biasing condition to measure word-line read retention voltage (WRRV) and **b** transfer curves to measure WRRV

zero, it means that the word line voltage can be boosted to above V_{DD} during the read operation. It is noteworthy that the word line voltage is slightly boosted during the read operation to strengthen the PG1 and PG2 to enhance the readability of the cell (This is considered as read-assist technique).

The large-scale metrics for the SRAM write noise margin are bit-line write trip voltage (BWTV) and word-line write trip voltage (WWTV). The BWTV and WWTV are measured under the biasing conditions in Figs. 8.14a and 8.15a, respectively. The biasing condition for BWTV is almost identical to that of the write operation in the SRAM cell. The major difference is that the BL' is driven to *GND* after both PGs are turned on. From the current versus voltage curve shown in Fig. 8.14b, we can define the BWTV as the minimum voltage where the measured current abruptly changes. If the BWTV is greater than 0, we can figure out that the SRAM cell can be written into, even if BL' is not fully driven to *GND*.

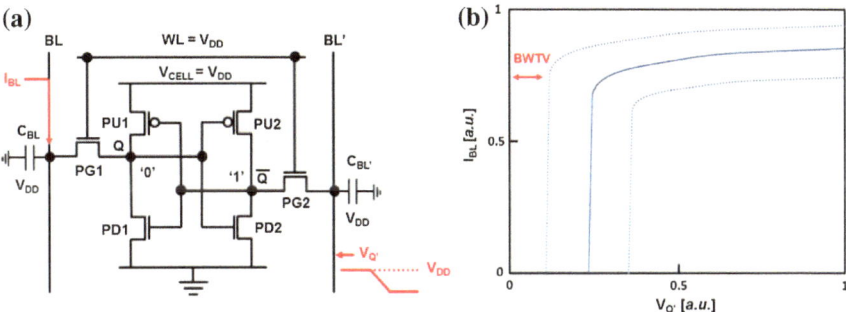

Fig. 8.14 **a** Biasing condition to measure bit-line write trip voltage (BWTV) and **b** transfer curves to measure BWTV

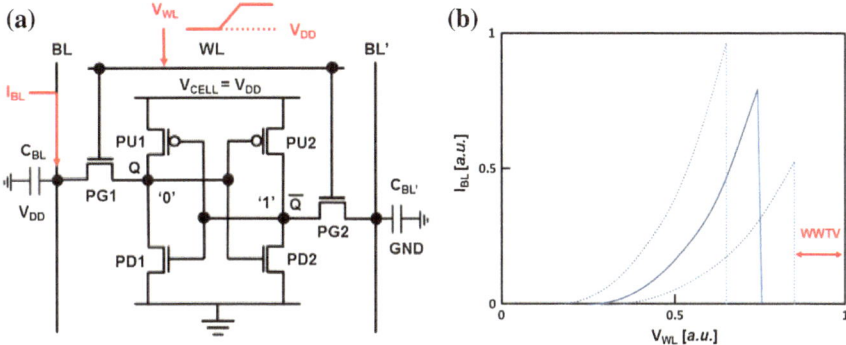

Fig. 8.15 **a** Biasing condition to measure word-line write trip voltage (WWTV) and **b** transfer curves to measure WWTV

The BWTV is quite different from the other large-scale margin metrics, but the word-line write trip voltage (WWTV) is almost the same as the WRRV. The major difference between WWTV and WRRV can be found in the biasing condition for BL and BL' (herein, note that word line voltage (V_{WL}) is swept from zero to V_{DD} for WWTV). The WWTV is defined as shown in Fig. 8.15b. Similar to the SRRV, if the WWTV is greater than 0, it means that the voltage applied at the word line during the write operation can be lowered without causing write failure.

The main difference between the conventional and large-scale metrics and the advantages of using the large-scale metrics over the conventional metrics are that the large-scale metrics provide maximum voltage at the word line as well as the minimum V_{cell} to avoid disrupting the fundamental operations of the SRAM cell.

8.4   SRAM Yield Estimation Techniques

This is the main section of this chapter. In this section, we will study how to estimate the yield of the SRAM cells when the process-induced random variations that we studied in the previous sections occur in the SRAM cells. There are a number of SRAM yield estimation methods, but we will focus on the yield estimation method based on TCAD (Technology Computer Aided Design) simulation with a compact model.

The full simulation for evaluating the SRAM cells can be performed by the mixed mode TCAD simulation technique. However, it consumes a lot of computation time to obtain a sufficient number of samples to make the statistical results reasonable [26]. Since the SRAM cells and the SRAM transistors are designed well enough to avoid any malfunction, the failure probability of a single SRAM cell is extremely low [27]. However, SRAM failure events always occur in SRAM arrays, because the number of bit cells on a single integrated chip (IC) is very large. Therefore, we focus on the yield estimation method based on a compact model, which can reduce the computation time and earn the sufficient number of sample data in a reasonably short time.

8.4.1   Compact Model for MOSFET

A semi-analytical current-voltage (*I-V*) model was derived for the fast and accurate estimation of the SRAM margin metrics by Carlson [21]. This compact model is similar to the SPICE model, which means that the *I-V* plot of the transistor can be simply re-configured with a few device performance parameters. In this model, the parameters for the *I-V* plot are determined using seven *I-V* targets—V_{THLIN} (linear threshold voltage), V_{THSAT} (saturate threshold voltage), I_{DSAT}, I_{DLIN}, I_{DHI}, I_{OFF} and I_{DLOW}, which are measured under the following biasing condition: (V_{GS}, V_{DS}) – (1.0 V, 1.0 V) for I_{DSAT}, (1.0 V, 0.1 V) for I_{DLIN}, (1.0 V, 0.5 V) for I_{DHI}, (0.0 V, 1.0 V) for I_{OFF}, and (0.5 V, 1.0 V) for I_{DLOW}.

From the compact model, we can fit the *I-V* data exported from the TCAD simulation to the re-configured *I-V* curve. The results are shown in Fig. 8.16. The

Fig. 8.16 Current-voltage (*I-V*) curves reconfigured by the compact model (see the *white-colored circles*), and the *I-V* curves extracted from the TCAD simulation (see the *solid lines*)

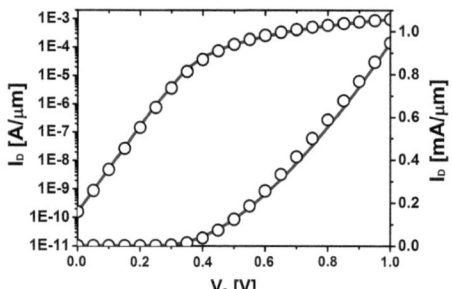

compact model can address physical phenomena like bulk charge effects and DIBL [28], velocity saturation [29], subthreshold conduction, and channel length modulation [30]. Hence, the impacts of process-induced random variations on the aforementioned physical phenomena can be analyzed using the compact model. The detailed information about the compact model can be found in [21].

8.4.2 Standard Monte Carlo Simulation

With the compact model in Sect. 8.4.1 [31], the voltage and current in the read/write operation of the SRAM cell can be determined using the Kirchhoff's Current Law (KCL) and Kirchhoff's Voltage Law (KVL) at each storage node. The exact values of the voltage/current at each node are determined by iteration. The initial values at each node are given for the read/write operation of the SRAM cell. Then, from the initial value, the KCL and KVL can find out the exact value of voltage/current for the given biasing conditions with iteration. Finally, the SRAM margin metrics can be quantitatively estimated.

The computation time for estimating the SRAM margin metrics with the compact model is reasonably short. However, if the capacity of the SRAM array is increased significantly, the amount of data for the SRAM margin metrics must be increased to estimate the metric to the six-sigma level [32]. However, as mentioned before, the failure events in a single SRAM cell are rare events. Hence, almost all of computation time is consumed in estimating trivial data for the margin metrics that lie within the three-sigma level [33] (see Fig. 8.17). Therefore, a statistical method must be used to reduce the computation time for estimating the margin metrics over/beyond six-sigma level.

Fig. 8.17 Measured WRRV in terms of sigma level

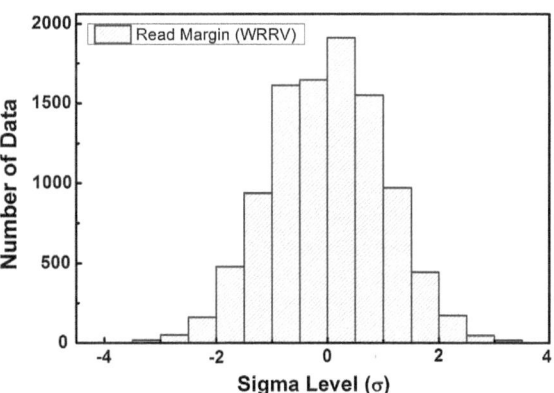

8.4.3   Worst Case Sampling Method

To overcome the limitation of the standard Monte Carlo simulation method (i.e., high computation time), many statistical methods to reduce the computation time for estimating the margin metrics over the six-sigma region were proposed [26]. In this section, the focus is on the worst case sampling method [27, 33], which is easy to understand without an in-depth background knowledge on statistics.

Based on the standard Monte Carlo simulations in Sect. 8.4.2, this method is used to seed data for the sampling method (herein, notice that LER, RDF, and WFV explained in the previous sections are implemented in the SRAM transistors, when the seed data is generated by TCAD simulation). Then, the number of samples (or sample size) is extended. In the process of extending the sample size, the seed data should follow a Gaussian distribution while the seed data do not strictly follow a Gaussian distribution. Hence, a power transformation [34] must be applied in order to transform the non-Gaussian distribution (i.e., the original seed data) into a Gaussian-like distribution. Since the Gaussian distribution can be characterized only by the mean and sigma of the distribution, the sample size is extended by the random number generation, which follows the Gaussian distribution extracted from the seed data. After the sample size extension, the data are inversely transformed to follow the previous distribution. The results are shown in Fig. 8.18.

After sample-size extension and inverse power transformation, the SRAM cells can be fabricated with the transistors characterized and re-built using the extended *I-V* data. In random sampling cases, 6 transistors from the whole *I-V* data are chosen. However, in the worst case sampling method, 6 transistors from the specific *I-V* data, which lie in the region where the sigma level is higher than the specific level, are chosen. Then, based on the KCL and KVL, the SRAM margin metrics are estimated with the chosen 6 transistors. Note that the SRAM margins at a much higher sigma region can be estimated with the worst case sampling method. Statistically signifi-cant levels of the read/write SRAM margin metrics (e.g., WRRV and WWTV), for a few cases of statistically significant levels of the sampling transistors under process-induced random variations, are explicitly summarized in Table 8.1.

Fig. 8.18 Estimated SRAM read and write metrics by worst case sampling method (vs. standard Monte Carlo simulation method)

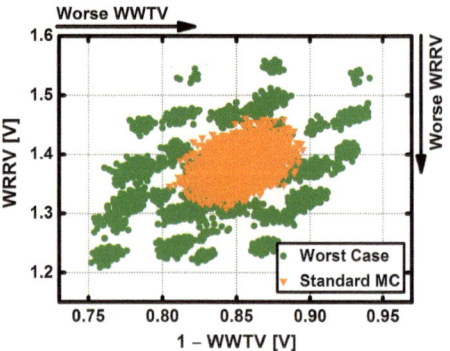

Table 8.1 Statistically significant levels of the read/write SRAM margin metrics (e.g., WRRV and WWTV) for a few cases of statistically significant levels of the sampling transistors under process-induced random variations

	$\pm 1\ \sigma$	$\pm 1.5\ \sigma$	$\pm 2\ \sigma$	$\pm 2.5\ \sigma$	$\pm 3\ \sigma$
WRRV	>4.5 σ	>5.1 σ	>5.8 σ	>6.4 σ	>8 σ
WWTV	>4.5 σ	>4.8 σ	>5.1 σ	>6.3 σ	>8 σ

References

1. Jan C-H, Al-amoody F, Chang H-Y, Chang T, Chen Y-W, Dias N, Hafez W, Ingerly D, Jang M, Karl E, Shi SK-Y, Komeyli K, Kilambi H, Kumar A, Byon K, Lee C-G, Lee J, Leo T, Liu P-C, Nidhi N, Olac-vaw R, Petersburg C, Phoa K, Prasad C, Quincy C, Ramaswamy R, Rana T, Rockford L, Subramaniam A, Tsai C, Vandervoorn P, Yang L, Zainuddin A, Bai P (2015) A 14 nm SoC platform technology featuring 2nd generation tri-gate transistors, 70 nm gate pitch, 52 nm metal pitch, and 0.0488 μm^2 SRAM cells, optimized for low power, high performance and high density SoC products. In: Proceedings symposium on VLSI technology, pp T12–T13
2. Agarwal K, Nassif S (2008) The impact of random device variation on SRAM cell stability in sub-90-nm CMOS technologies. IEEE Trans Very Large Scale Integr (VLSI) Syst 16(1): 86–97
3. Jianfeng L, Dornfeld DA (2001) Material removal mechanism in chemical mechanical polishing: theory and modeling. IEEE Trans Semicond Manuf 14(2):112–133
4. Asenov A (1998) Random dopant induced threshold voltage lowering and fluctuations in sub-0.1 μm MOSFET's: a 3-D "atomistic" simulation study. IEEE Trans Electron Devices 45 (12):2505–2513
5. Cathignol A, Cheng B, Chanemougame D, Brown AR, Rochereau R, Ghibaudo G, Asenov A (2008) Quantitative evaluation of statistical variability sources in a 45-nm technological node LP N-MOSFET. Electron Device Lett IEEE 29(6):609–611
6. Asenov A, Kaya S, Brown AR (2003) Intrinsic parameter fluctuations in decananometer MOSFETs introduced by gate line edge roughness. IEEE Trans Electron Devices 50(5): 1254–1260
7. Dadgour HF, Endo K, De VK, Banerjee K (2010) Grain-orientation induced work function variation in nanoscale metal-gate transistors—part I: modeling, analysis, and experimental validation. IEEE Trans Electron Devices 57(10):2504–2514
8. Brown AR, Idris NM, Watling JR, Asenov A (2010) Impact of metal gate granularity on threshold voltage variability: a full-scale three-dimensional statistical simulation study. Electron Device Lett IEEE 31(11):1199–1201
9. Asenov A, Kaya S, Davies JH, Saini S (2000) Oxide thickness variation induced threshold voltage fluctuations in decanano MOSFETs: a 3D density gradient simulation study. Superlattices Microstruct 28(5–6):507–515
10. Calhoun BH, Chandrakasan AP (2006) Static noise margin variation for sub-threshold SRAM in 65-nm CMOS. IEEE J Solid-State Circ 41(7):1673–1679
11. Dixit A, Anil KG, Baravelli E, Roussel P, Mercha A, Gustin C, Bamal M, Grossar E, Rooyackers R, Augendre E, Jurczak M, Biesemans S, De Meyer K (2006) Impact of stochastic mismatch on measured SRAM performance of FinFETs with resist/spacer-defined fins: role of line-edge-roughness. In: IEDM, 11–13 Dec 2006, pp 1–4
12. Liang-Teck P, Kun Q, Spanos, Costas J, Nikolic B (2009) Measurement and analysis of variability in 45 nm strained-Si CMOS technology. IEEE J Solid-State Circ 44(8):2233–2243

13. Kang K, Jeong H, Lee J, Jung S (2013) Comparative analysis of 1:1:2 and 1:2:2 FinFET SRAM bit-cell using assist circuit. In: International SoC design conference (ISOCC), pp 035–038
14. Pavlov A, Sachdev M (2008) CMOS SRAM circuit design and parametric test in nano-scaled technologies: process-aware SRAM design and test, vol 40. Springer Science & Business Media, Berlin
15. Calhoun BH, Chandrakasan AP (2007) A 256-kb 65-nm sub-threshold SRAM design for ultra-low-voltage operation. IEEE J Solid-State Circ 42(3):680–688
16. Kim K, Hamid M, Kaushik R (2008) A low-power SRAM using bit-line charge-recycling. IEEE J Solid-State Circ 43(2):446–459
17. Kim K, Kuang JB, Gebara FH, Ngo HC, Chuang C-T, Nowka KJ (2009) TCAD/physics-based analysis of high-density dual-BOX FD/SOI SRAM cell with improved stability. IEEE Trans Electron Devices 56(12):3033–3040
18. Irobi S, Al-Ars Z, Hamdioui S (2010) Bit line coupling memory tests for single-cell fails in SRAMs. In: VLSI test symposium (VTS), 2010 28th. IEEE
19. Shin C (2011) Advanced MOSFET designs and implications for SRAM scaling. Dr. thesis, electrical engineering and computer sciences, University of California, Berkeley
20. Ludwig T, Aller I, Gernhoefer V, Keinert J, Nowak E, Joshi RV, Mueller A, Tomaschko S (2003) FinFET technology for future microprocessors. Proceedings of IEEE SOI conference, pp 33–34
21. Carlson AE (2008) Device and circuit techniques for reducing variation in nanoscale SRAM. Dr. thesis, electrical engineering and computer sciences, University of California, Berkeley
22. Guo Z, Carlson A, Pang LT, Duong KT, Liu TJ, Nikolic B (2009) Large-scale SRAM variability characterization in 45 nm CMOS. IEEE J Solid-State Circ 44(11):3174–3192
23. Pilo H, Arsovski I, Batson K, Braceras G, Gabric J, Houle R, Lamphier S, Radens C, Seferagic A (2012) A 64 Mb SRAM in 32 nm high-k metal-gate SOI technology with 0.7 V operation enabled by stability, write-ability and read-ability enhancements. IEEE J Solid-State Circ 47(1):97–106
24. Hill C (1967) Definitions of noise margin in logic systems. Mullard Tech Commun 89:239–245
25. Grossar E, Stucchi M, Maex K, Dehaene W (2006) Read stability and write-ability analysis of SRAM cells for nanometer technologies. IEEE J Solid-State Circ 41(11):2577–2588
26. Kanj R, Joshi R, Nassif S (2006) Mixture importance sampling and its application to the analysis of SRAM designs in the presence of rare failure events. In: Proceedings on DAC, pp 69–72
27. Lee GS, Shin C (2015) Worst case sampling method to estimate the impact of random variation on static random access memory. IEEE Trans Electron Devices 62(6):1705–1709
28. Pierret RF (1996) Semiconductor device fundamentals, 2nd edn. Addison-Wesley, Reading
29. Streetman BG, Banerjee SK (2006) Solid state electronic devices, 6th edn. Pearson, Upper Saddle River
30. Neaman DA (2014) Semiconductor physics and devices, 4th edn. McGraw-Hill, New York
31. Nam H, Lee H, Lee GS, Park IJ, Shin C (2014) Analysis of random variations and variation-robust advanced device structures. J Semicond Technol Sci 14(1):8–22
32. Shin C, Damrongplasit N, Sun X, Tsukamoto Y, Nikolic YB, Liu Tsu-Jae King (2011) Performance and yield benefits of quasi-planar bulk CMOS technology for 6-T SRAM at the 22-nm node. IEEE Trans Electron Devices 58(7):1846–1854
33. Oh S, Jo J, Lee H, Lee GS, Park JD, Shin C (2015) Worst case sampling method with confidence ellipse for estimating the impact of random variation on static random access memory (SRAM). J Semicond Technol Sci 5(3):374–380
34. Box G, Cox D (1964) An analysis of transformations. J R Stat Soc Ser B (Methodological) 26(2):211–252